MW01602173

Master ChatGPT:
Build AI Assistants That Know Your Business

Learn to Build Your Very Own Custom AI Assistant That Writes, Thinks, and Responds Like You Would

Dan Wilson

Master ChatGPT: Build AI Assistants That Know Your Business

For permissions, bulk orders, or inquiries, please contact:
DataCurl East LLC
Cary, North Carolina
Web: www.datacurl.com

Cover design by Dan Wilson
Book design by Luke Kilpatrick

First edition: 2025

ISBN - Paperback: 979-8-9992400-0-2
ISBN - EPub: 979-8-9992400-1-9
ISBN - PDF: 979-8-9992400-2-6

Library of Congress Control Number: 2025915074
Printed in the United States of America

Dedication

I didn't mean to write a book. It's not like I tripped and fell, and a book came out. It's really the culmination of 25 years of systems and marketing experience, coupled with a desire to help, and the support of those who believe in me.

I dedicate this book to my wife, Shannon Scarlett. She's provided more support than a man has a right to expect, and her belief in my many crazy ideas inspires me.

I'd also like to thank Luke Kilpatrick for giving me the idea of writing this book and providing his help along the way. I would also like to thank him for telling me it would be *easy* (it wasn't). He's the author of the Mia Kingtide series of young adult books about the beauty, culture, and spirit of our coastal regions.

I'd like to thank Scott King and Charles Gantt for providing their inspiration along the way to fully embrace AI, both good and the bad. I've sharpened myself through many hours of bantering and experimenting with those guys.

I'd like to thank my cousin, Tiffany Miller, for always being ready to read and comment on early drafts of my work. Her effort made this book better.

I'd also like to thank you, the reader. To paraphrase an American philosopher, if I book falls in the forest, and no one is there to read it, is it really a book?

Important: How to use this book

This book walks you through setting up an AI assistant that knows your business. It's designed to be used sequentially.

Text Conventions

I've structured the book to use different fonts to help you know what type of content you are looking at.

Normal text in a serif font

`AI Prompt text in a larger Monospaced font`

`AI Output is in a medium monospaced font`

> *Tips and Reminders are in a callout style*

Worksheets

Throughout this book, you'll find templates and worksheets designed to help you customize your AI. I've kept the examples short, focused, and sometimes abbreviated, so you can grasp the concept without wading through walls of text. We all know attention spans aren't what they used to be.

When you're ready to dive deeper, check the **Appendix**. That's where you'll find the full templates, ready for you to adapt and use. Downloadable formats of the worksheets can be found at datacurl.com/books.

Extra Chapters

Look, there are plenty of marketing books out there. While I touch on marketing concepts when necessary, this book is really about building AI skills. To follow along, you'll need to make some informed choices along the way. One big one? Which messaging frameworks actually make sense for your business? What style will serve you best when your AI starts cranking out content?

That's why I've included a **Bonus Chapter** on 10 Messaging Frameworks that Work at the end. You may not need this, but it's there so you can get the basics you need to make smart decisions about how your AI represents you.

Contents

Foreword

Most marketers I meet share two feelings about generative AI: curiosity and whiplash.

You scroll LinkedIn, see some cool AI cheerleader gushing about automating everything, or another CEO claiming to be an *"AI first company"* and those who *"don't get it"* should work somewhere else.

You think, "I should be doing that." Five minutes later you open ChatGPT, type a question, and watch a flood of words that somehow feels both impressive and unusable.

Sound familiar? This book is for you.

I've been there. Standing at the edge of a technology that promises to change everything while secretly wondering if we're about to be replaced by a chatbot with better grammar.

Here's what I learned after hundreds of experiments, dozens of client projects, and more than a few AI-generated disasters: generative AI isn't here to steal your job. It's here to handle the parts of your job that drain your soul so you can focus on the parts that light you up.

I wrote these pages for marketers who know their craft but are new to working side-by-side with large language models. You don't need a computer science degree, a wall of prompts bookmarked, or a budget that scares finance. What you DO need is a practical roadmap that shows where AI fits, how to get quick wins, and how to keep control of your brand voice while the machine handles the heavy lifting.

My promise: every chapter gives you something you can try today without hiring extra staff or rewriting your tech stack.

Think of this book as a marketing funnel. You can jump to the stage that matches your needs:

Introduction: Fron Skeptic to Superuser

The introduction begins with the author's honest journey from skepticism to superuser. You'll see how he moved from AI frustration to AI fluency, and how you can too.

Part 1: How ChatGPT Really Works

Part 1 lays out how ChatGPT works under the hood, why it surprises you (in both good and bad ways), and how to think about AI not as a magic bullet but as a practical power tool. It reframes AI as a productivity amplifier, not a job thief.

Part 2: The Marketer's Guide to AI

Part 2 helps you shift from theory to action. It shows you how to actually use ChatGPT in marketing. From turning rough content into polished drafts to summarizing transcripts and remixing old assets into new formats, you'll be on the way to AI superstardom. This section also introduces prompt tactics and key mental models for working efficiently with AI, all grounded in a marketer's real-world workflow.

Part 3: Feed Your AI the Right Info

Part 3 is about building a solid foundation on which to operate from. You'll build your Company Brief, define your Ideal Customer Profiles (ICPs), map out Buyer Personas, and lock in your Brand Voice. Once complete, your AI assistant acts like part of your team, not a random intern guessing in the dark.

Part 4: Setting Up Your Custom ChatGPT Assistant

Part 4 walks you through building your own GPT that knows your business inside and out. You'll use the foundation from Part 3 to create a custom assistant that remembers your voice, understands your audience, and can help produce high-quality assets on demand.

Part 5: Core Messaging to Campaign Launch

Part 5 translates strategy into deliverables. You'll generate messaging pillars, landing pages, email sequences, and ad copy, all with the help of your AI assistant. This part shows how to prompt, refine, and package content quickly, while representing your business well.

Part 6: Beyond Email: Ad Copy That Actually Works

Part 6 gets tactical. You'll build Google Search and LinkedIn ads that resonate, experiment with tone and formats, and learn what converts. This is where the rubber meets the road: deploying copy into the wild with clarity, urgency, and specificity.

Part 7: Expand Beyond Campaigns

Part 7 opens up broader use cases: retrospectives, market research, podcast planning, case study production, and more. The goal here is to help you turn existing assets and insights into reusable content and smarter decisions without getting buried in grunt work.

Part 8: Deeper ChatGPT Training

Part 8 is where prompt-writing becomes a craft. You'll learn a proven structure for giving clear, effective instructions, see how to chain prompts together, and avoid common traps that derail AI output. It's your step-by-step to writing prompts that work on the first try, or at least get you close.

Part 9: Where To Go From Here

Part 9 wraps it all up. You'll explore the adjacent tools worth trying, sketch out your 90-day AI growth plan, and peek into the future of how AI will reshape marketing, business, and even scientific discovery. This is part reflection, part roadmap, part rally cry.

Appendix

Appendix materials include worksheets, templates, and a bonus chapter, "Hitchhiker's Guide to Messaging Frameworks." Whether you're stuck picking the right angle or need help structuring a company brief, this is where the rubber meets the spreadsheet.

The worksheets are also available in a downloadable format and optimized to be filled out and uploaded to your LLM of choice.

Read cover to cover, and you'll move from dabbling to delivering. Dip into a single section, and you'll still leave with a template, prompt, or tactic you can ship before your next status meeting.

Either way, AI rewards those who experiment and iterate.

A founder once told me, "I don't want to become an AI expert. I just want to stop feeling like I'm missing out." That's exactly what this book delivers. Not expertise for its own sake, or hype pushed out, so your dear author looks like a thought leader and futurist. I'm equipping you with skills for today.

Ready to turn that curiosity into campaigns that ship on time and on brand?

Turn the page.

INTRODUCTION

From Skeptic to Superuser

My Personal Breakthrough

I was drowning in half-finished projects.

Meet Dan (that's me), a mid-level product marketing manager juggling blog posts, newsletters, and campaign briefs. I knew what I wanted, but the constant start-stop on projects, mhessy notes, and shifting priorities were making me nuts.

Then I discovered how to configure ChatGPT to understand my business, products, and outputs. Within a month, my world changed:

- Cut writing and revising time by 75%
- Increased campaign volume without burning out
- Finally tackled my backlog of "almost done" tasks
- Found space to think strategically instead of reactively

Most importantly? I stopped feeling like I was drowning in half-finished projects that just needed one final push. One of the most important changes was my mindset.

I Went from Skeptic to Superuser (And You Can Too)

When AI first hit my news feed, I rolled my eyes. It's easy to be skeptical of new tech. I'd seen enough overpromises to know that most "breakthroughs" end up as more hype than help.

I figured AI tools would be fun to play with, but not something I could trust with real work. My projects are demanding and full of nuance. No two days are the same. How could a computer program handle that kind of variability?

Then I tried it.

First Impressions

My first real attempt? I fed ChatGPT a vague prompt about writing a blog post and waited for brilliance.

What I got felt… bad. Bland. Boring. No substance, lots of emotional and hyperbolic words that felt bolted on with no skill, story, or point.

While ChatGPT produced a lot of words formatted like a blog post, the content was unusable. I'd heard ChatGPT could write blog articles, and I suppose by the most liberal definition, it could. But I couldn't really *use* what it produced without heavy editing. What was the point?

I remember thinking, "Is this it? Is this the revolution?"

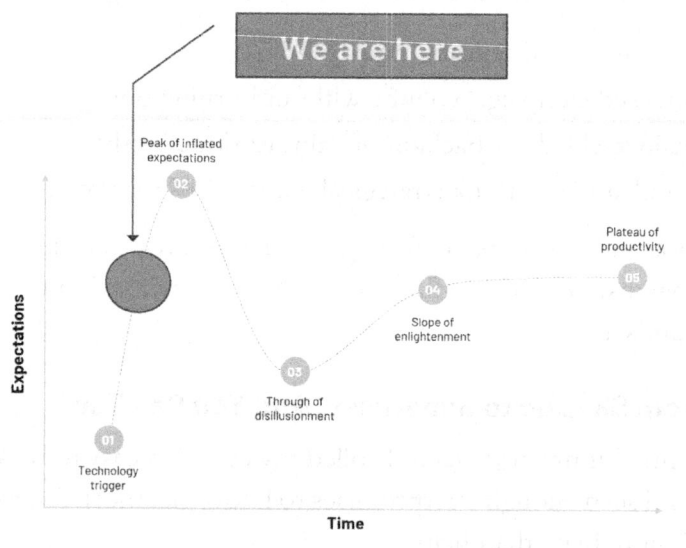

I bounced between extremes. One minute I'd hear AI was going to take over everything. The next, I'd see it was just a toy. I was right in the middle of what Gartner calls the "Peak of Inflated Expectations," and heading fast into the "Trough of Disillusionment."

I thought I was done with AI.

But then something shifted.

6

The Shift: Partnering with AI

The breakthrough came not from expecting AI to do my job, but from treating it like a partner that could help *me* do more of *my* job. Faster. Smarter. Better.

Everything changed when I stopped expecting AI to do it all and started using it where it could actually help. I took a hard look at my workflow and figured out what I had to own and where AI could give me leverage.

It all came together during a product launch. I work with software teams, so launches are a core part of what I do. They involve creating a story around new features, crafting content for different audiences, and driving outcomes like awareness and adoption.

These launches need more than just a list of new features. They need a message that lands. That means understanding what was built, why it matters, and who needs to hear about it. Framing that story is not something AI can do. That's where I step in.

Once the narrative is set, I move to execution. There are always follow-up pieces — landing pages, briefs, internal docs, and blog content. That's where AI becomes useful. It helps me move faster without starting from scratch every time.

My personal workflow begins with a narrated slide deck. It pushes me to get clear:

- What are we delivering?
- Who is it for?
- What problem does it solve?

There's something about the slide format that helps me organize my concepts and make them sharp. I can order and reorder them until it's just how I want.

Once I answer those questions, AI helps me cascade that story into every other format I need:

- Landing pages
- Campaign briefs
- Product marketing docs
- Blog posts
- Social post snippets
- Video scripts

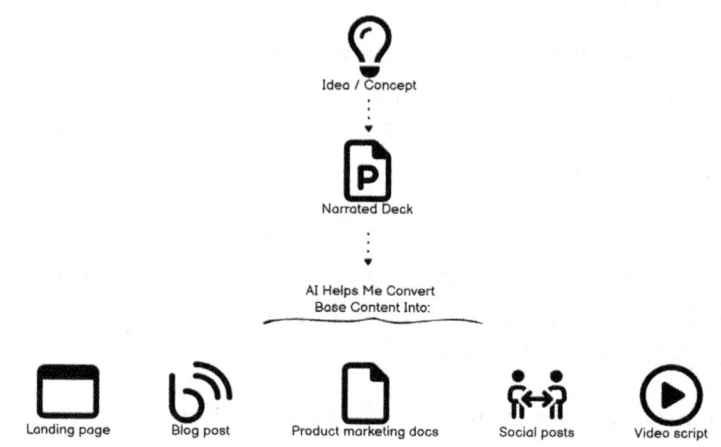

It was about doing more without giving up quality. I still had to think, revise, and guide the work. But I didn't have to carry the whole load.

I stopped expecting AI to do the job for me.

AI became a way to **augment my weaknesses** and **amplify my strengths** like storytelling, clarity, and strategy. I use the process that works best and most efficiently for me, then use AI to complete the rest of the work that I find boring, formulaic, or not my sweet spot.

That's the real power here.

Mapping the Journey to the Hype Cycle

Looking back, my experience follows the classic Gartner Hype Cycle:

- **Innovation Trigger**: I first heard about AI tools and got curious.
- **Peak of Inflated Expectations**: I tried it and expected magic.

- **Trough of Disillusionment**: It didn't blow me away, so I nearly walked away.
- **Slope of Enlightenment**: I figured out where it fits into my workflow.
- **Plateau of Productivity**: I use it every day, not as a crutch but as a multiplier.

If you're still in the early stages, take your time. Expect some disappointment but keep exploring. The payoff is in the shift, when you stop treating AI like a silver bullet and start using it like a power tool.

That's when things really take off.

Reconsider How You Work

The AI hype-sters love to sell a vision of AI "Doing All The Things™". It's just not there yet, and it may never be. They want to sell you their product, service, or a handful of magic beans by overpromising. There's a lot you can gain without needing to buy into the fantasy of AI doing it all. You'll need to open your mind and be self-reflective about how you work. What parts of your work do you do well, and which you aren't so good at?

It doesn't have to be some *Huge Thing* to make a difference. Let me give you a personal example that's a little embarrassing to me.

I never really learned how to type the way you're supposed to. You'd think after 25 years of writing code and content, I'd be better by now. When I do live demos or people watch me type, it gets even worse. I look like a clumsy toddler after too much apple juice.

See, we had typing class in high school. They taught Home Row/Touch Typing and to pass the class you had to hit 30 words per minute.

I'd been using computers for years, so I could already type 30 words per minute, but I wasn't a touch typist. My hands were too small when I first started, and I never slowed down and learned the correct way.

It made sense in my teenager mind to just skate through the class because I could already type 30 words per minute using a few fingers and a lot of backspacing. I've regretted that ever since.

Why does this matter?

My brain thinks in shapes, systems, and interfaces. If I stop to fix punctuation, spelling, or grammar as I go, the mental image can disappear. Once it's gone, it's hard to get it back. The key to staying in flow at the speed I like is capturing the ideas while they're still sharp. So, I'm always battling to maintain my mental design while battling backspacing. Then I get frustrated, lose my train of thought, you get the picture.

Now how does AI help? You've heard they are sticking computer chips in people's brains. I'm not there yet, but I do use AI as my trusty companion to stay in flow. Here's how:

Voice-to-text. Seriously.

Microsoft Word calls it Dictate. Google Docs calls it Voice Typing and hides it in the Tools menu. When I'm ready to get my thoughts out, I speak instead of type. I blab out my full idea without stopping to revise. Of course I end up with word soup, bad grammar, misspellings, and a run-on sentence to end all run-on sentences because the designers behind these voice-to-text tools clearly didn't grow up around southern accents.

The transcripts aren't accurate, and they're never formatted in a usable way. If I had to clean those up by hand, it would ruin both the fun and the productivity.

This is where our friend ChatGPT comes in. It has no trouble understanding southern. I run the voice draft through it, and it turns my rambling phrases into clean, readable, structured content.

I still do a final pass, but the heavy lifting is already done. I've captured the idea without wrestling with typos or getting stuck on grammar. It keeps me working at the speed of thought, not the speed of my typing.

This is one small way that doesn't show up as a "Game Changing AI Use Case", but it makes me faster, better, and happier at work.

Reflection

As you think about integrating AI into your work, focus on how it supports the way *you* think, work, and create. Everyone's brain works a little differently. The real power of these tools is how they can bend to fit your flow.

Try this: Take a few minutes to reflect on these questions:

- What parts of your workflow create the most friction or distraction?
- What parts of your job do you ignore because you just don't want to do it?
- When are you most "in flow," and what helps you stay there?
- What do you do out of habit, even though it's not the most efficient way?
- If you had a magic wand, what would be different about how you work?
- Where could AI handle the setup, draft, or grunt work so you can focus on judgment and creativity?

Answer honestly. The more awareness you build, the easier it becomes to design a system that works for you.

PART 1
HOW CHATGPT REALLY WORKS

Before we get too far into this book, it's important you learn a few key details about how this technology works. In later chapters, we go deeper, but I want to give you these concepts early, in case you are the kind of person to skip around. You'll thank me later.

CHAPTER 1

A practical primer for using the tool wisely

Here's the thing about ChatGPT most people get wrong: they think it's thinking. ChatGPT isn't smart. It's just really, REALLY good at guessing what comes next.

It's All About Prediction

ChatGPT is like that friend who's read everything and has an opinion about everything. During its training, this system consumed massive chunks of internet. Books, forums, research papers, code repositories, it vacuumed up all things digital and used complex math to find patterns, relationships, and other super geeky things. By analyzing which words tend to follow other words, how sentences typically flow together, and what knowledge connects to other knowledge, it can produce what appears to be new, novel content.

See, when you type a prompt, ChatGPT doesn't think or search through some vast knowledge base. It calculates probabilities. "Given these words, what's most likely to come next?" Then it commits to that choice and repeats the process, word by word (actually tokens, but we don't care), until it hits a length limit or decides it's done.

Why does it sound so brilliant? Well, human writing itself often follows patterns. We use similar phrases, build arguments the same ways, tell stories with familiar structures. ChatGPT absorbed millions of examples of how we communicate and can mimic this like a digital parrot who consumed too much coffee.

Why This Matters for Your Marketing

Because the model optimizes for likelihood, it remixes ideas in ways you might never think of. That's gold for brainstorming campaigns, testing copy variations, or cranking out headline options. There's rarely one "right" answer in creative work anyway.

However, there are downsides.

The Memory Problem Nobody Talks About

Ever seen a carnival coin-pusher game? The one where you drop a coin in, and it nudges other coins forward until some fall off the edge? ChatGPT's memory works exactly like this.

Every conversation has a fixed-size "context window." As you add new information, the oldest and least important details get pushed out and disappear forever. The model won't warn you when this happens. It just forgets.

I've poured hours into detailed briefs, only to watch ChatGPT gradually lose track of my brand voice, target audience, or some other detail. The output gets worse and worse. Turns out, the longer the conversation, the more likely you are to hit this invisible wall.

Most businesses treat ChatGPT like it has perfect recall. Few prepare for its goldfish memory.

Why ChatGPT Keeps Surprising You

Traditional software behaves like a vending machine. Same input, same output, every time. ChatGPT deliberately doesn't work this way.

The model's variability explains why identical prompts produce assorted responses. It's supposed to do that. In marketing, where fresh perspectives win, the quirk is a feature. In software circles, where predictability rules, it feels like a flaw. Keep this in mind: push for inspiration, not perfect repetition, or the tool will seem uncooperative.

The Confidence Game

Here's ChatGPT's biggest weakness: it never admits it doesn't know something.

Because the system must always generate the "next most likely token," it will confidently invent citations, statistics, or product features when the real information isn't available. It's an inevitable consequence of how large language models work.

I call this the "confident liar" problem. ChatGPT doesn't know when it's making things up because it doesn't distinguish between facts and probable-sounding text. This issue has been improving over time, but it'll never be perfect.

When Humans Must Stay in Control

Statistical guesses break down when stakes get high or situations lack clear precedent. Medical advice, legal compliance, financial decisions, even your market strategy, these areas require judgment that sits outside token probabilities. If the model has never encountered a specific scenario, it'll improvise based on vaguely similar patterns. The responsibility to know the difference lies with us. You'll get better at this with practice. Let's move on.

PART 2

THE MARKETER'S GUIDE TO AI

CHAPTER 2

Why AI Isn't Coming for Your Job

If talk of "AI-generated content" or "automated workflows" has made you uneasy, you're not alone. But here's the deal: AI isn't here to take your job

We've seen this movie before. When the typewriter showed up, writers didn't vanish. When Photoshop launched, designers didn't throw in the towel. AI is the next wave. Like past tools, it gives you leverage, not a pink slip.

Let's be honest: There's a lot of hype around AI, and a lot of fear baked into that hype.

As the cost of doing great work drops, we don't run out of work, do we? We raise the bar. Expectations rise. If every image had to be crafted by a human designer, we accepted the limits of time and cost. But if images can be generated in seconds with AI, the market starts to expect visual content everywhere.

The Rise of AI in Marketing

According to McKinsey, over 75% of marketers are already using AI in their work. Sometimes without even realizing it.

It's in your inbox, your website analytics, your ad campaigns, and increasingly, your creative toolkit. From chatbots to automated ad bidding, the tools have arrived. They're getting smarter and more prevalent by the day. I'm expecting Amazon to show me an AI-powered hairdryer any day now.

And today's AI isn't reserved for data scientists. It's built for you. Well, mostly. It's still a bit rough around the edges. I'll show you how to work through some of those cases in later chapters.

A Productivity Booster, Not a Threat

The real magic of AI isn't that it writes like a human. It's that it works like a machine. It never tires, never runs out of energy, and doesn't care how many drafts it generates. It doesn't call in sick, ask for more vacation days, or even microwave smelly meals in the office microwave.

AI gives you leverage. Want to test 12 versions of a subject line instead of 3? Done. Want to reword a LinkedIn post for a different audience segment? Easy. AI handles repetitive drafting, editing, and formatting, but not creative judgement. That's still your domain, you just have more time to focus on it.

Avoiding the Hype Trap

Now, a quick warning: There's a lot of hype out there. Everyone wants to sell you something, AI products, news article clicks, even business books (*/ me ducks*).

AI is impressive, but it's not magic. Behind the curtain, it's a word prediction engine with a smooth voice and a knack for metaphors. It doesn't think, feel, or create. It's an illusion, and a very good one.

It doesn't understand your unique brand context or audience pain points. It doesn't replace empathy, ethics, or judgment.

AI sings a siren song: "Look how much I've written! Just copy me!" And yeah, it's a seductive pitch. No editing, no stress, done before lunch. But that song often leads you straight into the rocks. Sounding smart isn't the same as being smart.

The marketers who get the most out of AI aren't the ones who treat it like a magic wand. They're the ones who treat it like a teammate, maybe even a slightly incompetent teammate.

Getting to Know AI's Role in Marketing

> ### Note:
>
> *The advice and principles in this book work for pretty much any generative AI platform. I use ChatGPT for this book because it gets too confusing switching back and forth between platforms and ChatGPT makes it super easy to "train" your marketing assistant.*

What ChatGPT Brings to the Table

At its core, ChatGPT is a text-based AI assistant. You give it instructions (called "prompts"), and it generates output.

AI is amazing at:

- Brainstorming content ideas
- Drafting first versions of text
- Rewriting content in a new tone or format
- Summarizing long documents or transcripts

But it's not great at:

- Making nuanced brand or legal decisions
- Handling confidential or proprietary strategy
- Writing final drafts without review

Here's the part most marketers love: speed. AI breaks the mold of repetitive marketing templates. If you love to use a good template to get your flow and focus going, you're really going to love using AI to quickly personalize content by industry, persona, or funnel stage.

Checklist: Are You Ready for AI in Your Workflow?

Ask yourself:

- Do you suspect you spend too much time on your tasks?
- Would more variants help your campaigns perform?

- Are you open to editing, not writing from scratch?
- Are you curious enough to experiment?
- Am I willing to treat AI output with the skepticism it deserves?

If you said yes to the above, you're ready.

CHAPTER 3

How To Use Your AI Marketing Assistant

This book walks you through building your own AI-powered marketing sidekick. By the end, you'll have a custom assistant that writes like you, understands your audience, and works in your style.

To get you thinking, let's go over where you might use this assistant.

AI Supercharges Your Funnel

Marketing is about identifying groups of potential prospects then guiding them from curiosity to commitment, and ideally to loyalty.

Here are ways AI supports your strategy.

The Full-Funnel AI Playbook

AI can plug in at each stage:

- **Top-of-funnel:** Brainstorm headlines, create SEO snippets, write blog drafts, generate social post variants
- **Middle-of-funnel:** Personalize email campaigns, segment audiences, create nurturing flows triggered by behavior
- **Bottom-of-funnel and post-sale:** Analyze customer success data, surface upsell opportunities, write win-back sequences, summarize support transcripts

Plus, AI tools can do deep research now. They can crawl the web, gather facts, and build useful summaries with links you can check yourself. Once you get a feel for it, you'll start seeing chances to use it all over your workflow.

> *Tip:*
>
> *AI is infrastructure, not intelligence. Developing smart marketing strategy still belongs to humans. That's the fun part (and the hard part). If you don't have a clear direction, AI will just help you get lost faster.*

Work with What You've Got

Before creating anything new, pause. You're probably sitting on a goldmine of half-finished, forgotten, or lightly used materials like case studies, sales notes, product decks, transcripts, even customer emails. This is where AI can help you move from "raw" to "ready."

AI can help you clean them up, group them, and turn them into something sharper. You might go from five messy decks to one solid fact sheet, or from ten case studies to a clear profile of your ideal customer.

The trick is to stop thinking of these as finished or useless. They're building blocks. Let the AI help you stack them into something better.

Here's a concrete example:

1. Gather a few messy assets: a customer quote, a demo transcript, maybe a campaign recap.

2. Grab a case study you like (from your own library or a competitor).

3. Give both to ChatGPT with this prompt:

You're an expert B2B marketer. Here's a sample case study. Based on the format and tone, create a prompt I can use to generate a new case study using this raw customer input.

---

4. Feed in your rough material and start iterating.

5. Once you have the new case study, turn it into a blog. By the end of this book, you'll have enough experience to do this in a few minutes.

It seems obvious once you do it, but if you aren't sure of the exact prompt to use to hit your goal, you ask ChatGPT for the detailed prompt.

Promptception

Yes, that's right. You don't have to be a prompt expert. AI can write great AI prompts. It's a bit like the movie Inception.

Whenever you're stuck or need a jump start for a good prompt, ask the AI tool to write the prompt it wants.

It sounds strange but think about it. What tool is more qualified to shape the question than the one answering it?

Tip:

Ask AI to build the prompt for you.

Freedom to Discard

Given all that, we don't have to worry about perfection. Let the AI take the first swing, then decide if it's worth keeping. There's no sunk cost and no attached ego. That's the upside of low-cost output. You're free to focus on what's actually good, without letting the amount of effort color your judgment.

Oh, and restrain yourself from altering or polishing the output until you have something good. I know, it's hard to not correct issues you see, but trust me, training yourself to wait is worth it.

Tip:

If you polish too early, you refine content that ends up in the trash. If you focus on getting the perfect prompt, you spend time writing prompts and not producing content.

Sometimes the best move is to stop fiddling and just edit the output yourself.

Tip:

If you catch yourself rewriting the prompt five times, ask yourself, would a manual edit have been faster? (I do this all the time.)

Tidy the Mess, Spin Gold

Now that you've thought about your pile of content, let's talk about derivative content.

- Combine 3 old blog posts into a whitepaper or ebook draft
- Take an outline and slide deck and spin up a webinar script
- Extract social quotes or email intros from a webinar transcript

Repurposing approved content is one of the fastest ways to scale. If something's already been published, it's been through review, feedback, and editing. That makes it a strong input for AI.

I like to start with legacy pieces and turn them into a core asset. Then I use AI to convert the ideas and material in the core asset into emails, landing pages, sales enablement material, or whatever content items I need.

We want speed, we demand accuracy. Starting from a known-good position makes all the difference.

AI Doesn't Just Do Words

If you only used AI to generate copy, you're missing out. Multimedia no longer has to be a bottleneck. You don't have to be a designer or video editor to make creative things, and AI helps you move faster in channels that used to be slow, technical, or expensive.

Modern AI tools can:

- Create product mockups and illustrations
- Generate voiceovers from text
- Edit video clips or generate captions
- Build presentation decks from bullet points

Ideas:

- Marketers can test thumbnail images and ad copy faster
- Creatives can explore different concept directions quickly
- Social teams can prep clips in minutes, not hours

Just be aware that quality is in flux, and while we might be in love with "cool tech", our audiences may respond differently. I personally cringe when AI avatars pretend to speak. There's something off that gives me ick.

Tip:

Stay on brand. Sometimes cool, isn't cool.

How I use AI day-to-day

I drop five sales call transcripts into ChatGPT and ask it to extract customer pain points, objections, and recurring themes. From those, I draft campaign angles.

I feed in a few blog articles and ask it to generate a first-pass ebook. Or I do it in reverse—break an ebook into a multi-post blog series.

I upload a video transcript and request:

- Social posts
- Email summaries
- A landing page draft
- Key quotes and callouts

Even this book started as something else. I published a free video course on Udemy, but the strict 2-hour time limit meant I had to cut a lot of ideas I wanted to include. Rather than re-record everything and charge for the course, I decided to write a book.

I grabbed the transcripts and used ChatGPT to help clean them up. I reorganized the content, filled in the missing ideas, and rewrote parts that didn't work on the page. There was *plenty* of editing involved. I wrote and rewrote every paragraph (many times). The AI helped, but it was up to me to think through the narrative and make it feel like something worth reading.

What really saved the most time was asking ChatGPT to convert the perspective from "video course instructor" to "book author". That one step gave me a solid foundation to write. It turned the work into a creative and strategic set of tasks I'm more suited to and helped me skip boring perspective rewrites. It also provided a clean structure. Definitely helpful.

Sharpen as You Go

Shipping a campaign is step one. After that, it's all about optimization. And that gets messy. Ten subject lines. Four versions of copy. Performance summaries. Asset updates. Then another round of edits. It's not that the work is hard, but with all the other work and priorities on your plate, the constant context switching gets annoying.

AI can help with:

- Drafting dozens of subject line variants fast

- Rewriting CTAs in different tones and styles
- Summarizing A/B test results into one-line takeaways
- Segmenting performance by time, region, persona, or channel
- Suggesting next steps or adjustments based on results
- Pulling quotes from customer feedback to use in wrap-up decks

Use AI to compress the heavy cognitive lift of analysis and revision so you can focus on decision-making. The more the machines handle grunt work, the more you focus on what moves the needle.

I know this can seem like a lot. It might be best to revisit these first few chapters in a few months once you've gotten the hang of AI in your workflow. You might be surprised how different some of these ideas sound then.

Tip:

Come back to this book later, especially these first few chapters. It'll spark new ideas and learning.

CHAPTER 4

Before We Get to Building

This book teaches rather than consults. I've simplified concepts to make them clear and actionable. Real situations have more nuance, so use these frameworks as starting points. Adapt them to your context and trust your judgment.

Want to debate these frameworks? I'm game. Just leave a 5-star review first, then we can hash it out in the comments.

Staying Smart and Safe with AI

Fast doesn't always mean safe. When you're moving quickly with AI, it's easy to overlook issues hiding in plain sight. One wrong sentence, one bad fact, or one off-brand message can cause more damage than the speed was worth.

Remind me one day to tell you about the huge campaign I had to cancel and forgot one automated email. VPs were involved. There was potential legal action. And this was self-inflicted. No AI was used in the making of that catastrophe.

See, AI is a powerful tool, but it doesn't come with a safety switch. You still must steer.

Let's talk about what to watch out for, because you are responsible for what you publish.

Warnings

Watch out for:

- Copyright misuse or repurposing of third-party work
- Hallucinated facts or made-up numbers
- Outdated information that sounds current but isn't
- Fake references or citations that look real but don't exist

- Messaging that strays off-brand or feels tone-deaf
- Conflicting claims or numbers across related content pieces
- Repeating the same phrasing or structure across assets without realizing it
- Generating content that violates accessibility or compliance guidelines
- Incorrect use of brand voice or tone in regulated industries
- Biased content pulled from flawed training data

And the list goes on. A few habits I recommend:

- Verify facts before using them in research or content
- Don't label something "human-written" unless it is
- Read outputs carefully for bias, awkwardness, or outdated phrasing
- Avoid uploading anything confidential
- Start with accurate, vetted content
- Run AI drafts side-by-side with your top-performing or officially approved content
- Before hitting publish, run the content through a checklist:
  - o Are all facts sourced?
  - o Is the tone aligned?
  - o Are key claims consistent with other assets?
  - o Are accessibility guidelines met?

Oh, and also important:

- Upload brand-approved language, product positioning docs, and past campaigns into your AI workspace.

We'll be doing this in later chapters, so stay tuned.

Next up: Let's get hands-on and walk through the ChatGPT interface.

Getting Started with ChatGPT for Business

Before we build your custom marketing assistant, let's make sure you're set up with the right tools.

We'll be using ChatGPT, specifically the version available with the **ChatGPT Plus** plan. If that's new territory for you, no worries. This chapter gets you up to speed.

Choose the Right Plan

To follow along with everything in this book, ChatGPT Plus is your best bet. It gives you access to:

- **GPT-4**, which generates better content and handles instructions more accurately.
- **Custom GPTs**, which let you build your own AI assistant that remembers your preferences.

If you're on the free plan, you can do many of the exercises, but you may need to manually re-paste instructions more often.

Get Familiar with the Interface

The ChatGPT layout is clean and simple:

- **Left Sidebar:** Your past chats and, if you're a Plus user, your custom GPTs.
- **Main Chat Window:** Where all the action happens.

Key features you'll use throughout this book:

- **Custom Instructions / Custom GPTs:** These let you give the AI a consistent voice, purpose, and knowledge base. We'll build one in Chapter 7.
- **File Uploads:** Handy for sharing docs like product briefs or persona sheets.
- **Prompting in Chat:** This is where you type commands, questions, or tasks. Don't worry, we'll walk through how to write great prompts.

Take a second to log in and explore. If you see "My GPTs" in the sidebar, you're good to go.

Set Your Expectations

Here's the deal with AI: what you put in matters. Like we say in computer science:

Tip:

Garbage in, garbage out.

ChatGPT doesn't know what you mean. It only knows what you say. Clear input leads to clear output. Vague input gives you a mess.

Play It Safe with Sensitive Info

One last thing: be smart about data privacy. OpenAI may use your chat data to improve the model, unless:

- You turn that off in settings.
- You use the Enterprise version.
- You upload files via API or a custom GPT (these aren't used for training).

AI is still a young technology, and even its creators don't fully understand everything it can do. That's why you need to handle it with care. Unlike a website, where you can remove a mistake, content that makes its way into a language model may be impossible to take back.

Be smart. Use judgment. Don't feed it sensitive or personally identifiable information. Your job might not depend on AI, but it does depend on how wisely you use it.

> *Tip:*
>
> *Don't paste in confidential or personally identifiable information. If you need to reference real data, tweak it slightly or use examples.*

With these basics locked in, you're ready to start teaching ChatGPT about your business. In the next chapter, we'll build your Company Brief. The brief is a quick, powerful foundational document to make sure your assistant knows who you are, what you do, and why it matters.

10 Prompts To Play With

Get familiar with ChatGPT by trying these prompts:

1. Make It Make Sense (Gen Z Edition)

Your boss sent you a 17-page whitepaper. You just want the gist.

Prompt:

```
Act like a Gen-Z language model. Summarize this article in
3 bullet points. [article text or link]
```

Outcome: Three sharp bullet points with less jargon, more clarity, and maybe a little sass.

2. Headlines Without the Existential Crisis

When you've stared at the screen for 20 minutes and written: "Welcome to our website."

Prompt:

```
Write 3 headline options for this landing page [landing
page copy]
```

Outcome: Three headline ideas you didn't have to fight your inner critic to produce.

3. Pitch It Like a Local

Because small business owners don't speak "enterprise SaaS enablement."

Prompt:

```
What are some ways to position this product for small
business owners [short product description]
```

Outcome: Positioning that sounds more like a helpful neighbor, less like a press release.

4. One Post, Many Lives

Your blog deserves better than a quiet life in the archives.

Prompt:

```
Turn this blog post into a LinkedIn post and email intro
[blog post text]
```

Outcome: More reach, less rework. Your content now multitasks like a marketer on a deadline.

5. Subject Line Spark Generator

Because "Quick update" isn't fooling anyone.

Prompt:

```
Suggest 5 subject lines for this email [email content]
```

Outcome: Better open rates and fewer groans from your subscribers.

6. Explain It to a Golden Retriever

If your feature pitch requires a dictionary, it's time to simplify.

Prompt:

```
Explain this feature in simple language for a non-technical
buyer [feature description or product copy]
```

Outcome: Clear messaging that even your mom could understand. Or your CFO.

7. Campaign Gut Check

Before you spend five figures promoting it, maybe see if it's any good.

Prompt:

```
List the pros and cons of this campaign idea [short campaign
description]
```

Outcome: Instant perspective. A quick filter to decide whether to build it or bin it.

8. Update Without Overselling

Your product shipped a new button. It's cool but maybe don't shout.

Prompt:

```
Write a product update announcement that sounds excited but
not too salesy [new feature or release details]
```

Outcome: An update that sounds confident, not cringey. Customers stay informed, not annoyed.

9. Landing Page Last Looks

Because "Looks fine to me" is not a QA process.

Prompt:

```
What's missing from this landing page [landing page copy]
```

Outcome: Suggestions for improving clarity, flow, or persuasion before your boss sees it.

10. Next-Step Navigator

You finished the campaign brief. Now what?

Prompt:

```
Based on this campaign brief, what should I create next
[campaign brief or summary]
```

Outcome: Logical next content pieces to keep momentum going and avoid blank-page paralysis.

PART 3

FEED YOUR AI THE RIGHT INFO

CHAPTER 5

Building Your Company Brief

Every assistant needs proper context. Before ChatGPT can create content for you, it needs to understand your company, market, and products. This starts with a "Company Brief" – a concise overview of your key business facts and messages. Think of it as the one-pager you'd give a new marketing hire.

Tip:

Customize the Minimum Viable Company Brief Worksheet in Part 1 of the Appendix for this exercise.

What is a Company Brief?

It answers fundamental questions about your business: who you are, what you offer, who you serve, and why you're different. Not a business plan, but a minimum viable brief with just enough information for marketing purposes.

Your brief should answer:

- Who are you? A one-liner about your company (industry, product category)
- What do you offer? Your main product/service and its value
- Who do you serve? Your target customers or market segments
- When do they need you? Situations when customers seek your solution
- How do you help? Specific problems you solve or benefits you deliver
- Why does it matter? – The bigger impact behind your product (your "why")
- Messaging Pillars – Key points for foundational messaging

- Use cases – Examples of how customers use your product
- Competitors – Alternative solutions or rivals in your market

Take time to write bullets or short paragraphs for each point. Keep it factual and clear.

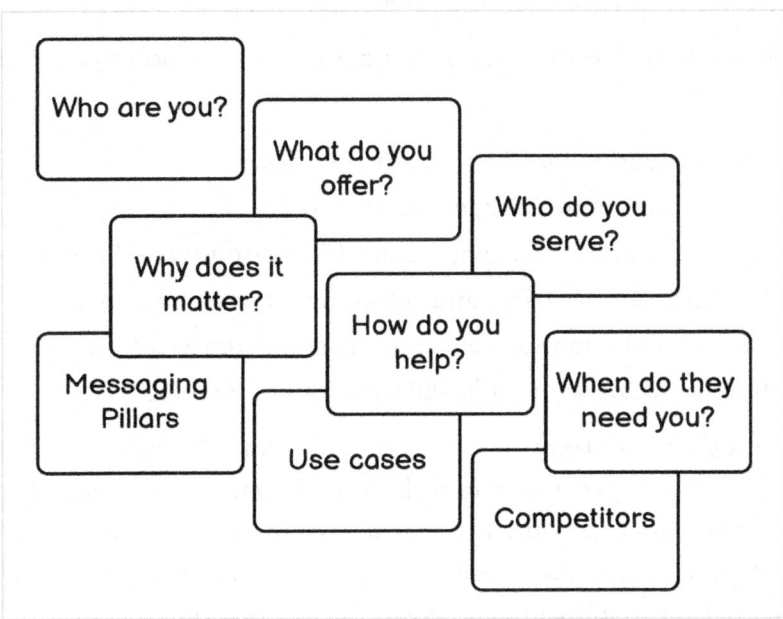

Example Brief: AnyQuest.ai

Who are you for? – Companies with $1M–$1B in revenue that have started using AI tools internally but need help scaling AI workflows across teams. Common buyers are in SaaS, logistics, professional services, or other mid-market to lower-enterprise orgs. Ideal personas include: CIOs, Heads of Ops, Innovation Leaders, AI Champions, and skeptical CFOs or IT stakeholders.

What do we offer? – A no-code/low-code AI enablement platform that lets organizations build, deploy, and scale multi-step AI agents that integrate with their existing tools. It simplifies AI adoption by eliminating manual prompt work, enabling agent-to-agent workflows, and centralizing knowledge via retrieval-augmented generation.

When do they need us? – When they've experimented with ChatGPT or AI tools, but workflows are manual, inconsistent, or siloed. Often under pressure from leadership to make AI more productive and want to reduce reliance on a few technical experts. Buying is triggered by internal pilots stalling, inconsistent output, or unmet AI scaling goals.

How do we help them? – The specific problems you solve or benefits you deliver.

- Visual agent builder (multi-step workflows)
- Integrations with Slack, email, and APIs
- Retrieval-augmented generation (RAG) for internal knowledge reuse
- Hosting, summarizing, and searching internal content
- Cross-agent communication to maintain consistent outputs
- Extend LLMs with tools like web search, scraping, etc.

Why does it matter? – Customers want to move beyond AI experimentation into consistent, scalable, and secure automation. They want to empower teams (not just experts) to use AI, reduce manual effort, and speed up research, content generation, or internal processes — all while ensuring reliable, compliant, and cost-effective outcomes.

Messaging Pillars: Key points we should use for foundational messaging.

- No-code AI workflows for operations, support, and product teams.
- Go from idea to deployed AI workflow in under an hour.
- Connect to your tools, databases, content, systems, and APIs.
- Build Agents that understand your data, customers, and processes
- Turn AI into a process you can trust and scale.
- Build once, reuse everywhere

Use cases: A few examples of how customers use our product.

- Account Research Bot – Pulls CRM + web data to generate sales briefs.

- Competitive Intel Tracker – Monitors and summarizes competitor activity.
- Compliance Checker – Reviews docs against policies or standards.
- Contract Analyzer – Extracts key terms, risks, and dates from PDFs.
- Customer Health Snapshot – Builds QBR decks from usage + notes.
- Escalation Triage Agent – Routes and prioritizes support issues.
- Internal AI Helpdesk – Answers employee questions using company docs.
- Objection Handler – Surfaces rebuttals during live or recorded calls.
- Onboarding Assistant – Guides new hires or customers through checklists.
- Weekly Report Generator – Summarizes dashboards + emails into updates.

Competitors: No-code/low-code AI agents for internal ops:

- Reka
- Kleon
- Cognosys
- MindStudio by YouAI

Loading your Brief into ChatGPT

ChatGPT will "remember" this information for the duration of the chat and we rely on this behavior to draft additional documents in the next few chapters. To make our assistant useful for us, we'll need to ensure ChatGPT doesn't "forget", there are two options for this.

1. With ChatGPT Plus – we upload files to a Custom GPT (we'll cover this fully in Chapter 9)
2. Without ChatGPT Plus, you must copy-paste all important context into ChatGPT for each new chat.

The 2nd option quickly gets arduous and error prone. That's why we need ChatGPT Plus. The magic of ChatGPT Plus is in creating specialty AI agents that'll help you in very specific ways. Once you get comfortable with this process, you'll create several different AI Assistants, each tuned to a certain need or business function.

Test your brief with this prompt:

```
Here's a company brief. Summarize the product, target
audience, core pain points, main benefits, and key
differentiators. Let me know if anything feels unclear or
missing.
```

This serves as a comprehension test. If ChatGPT misses something important, tighten your brief.

By doing this, you:

1. Load your company context into the conversation
2. Verify the information is correctly understood

Try this: Fix any misunderstandings now, even small ones. It's easier to correct the input than to fix multiple content pieces later because the AI went in the wrong direction.

With your company foundation in place, it's time to define who you're trying to reach. Next up: creating your Ideal Customer Profile.

CHAPTER 6

Defining Your Ideal Customer Profile

Generic messaging feels safe but gets ignored. A message aimed at "everyone" resonates with no one. It's like cooking without spices – looks fine but tastes bland.

Your Ideal Customer Profile (ICP) focuses your marketing on best-fit customers – those who'll see real value in your offering and actually buy. Start with who benefits most and is easiest to reach.

For B2B, an ICP describes a model company. For B2C, it describes an ideal consumer group. We'll focus on B2B context here (since our example "AnyQuest" is B2B), but you can adapt the principles to B2C.

Tip:

Customize the Ideal Customer Profile (ICP) Worksheet in Part 1 of the Appendix for this exercise.

What to include in an ICP

A solid ICP covers several aspects of your target company:

Firmographics: Basic identifiers

- Industry: Sectors or verticals
- Company size: Revenue range, employee count, locations
- Geography: Where they operate
- Ownership/structure: Private, public, VC-funded, etc.
- Growth stage: Startup, scale-up, enterprise, etc.

Technographics: Their tech profile

- Current tech stack: Software they use (especially relevant to your product)
- Tech maturity: Early adopters or more conservative

Business Model & Structure: How they operate

- B2B/B2C/B2B2C: Who their customers are
- Go-to-market motion: Sales-led, product-led, channel partners, etc.
- Decision-making: Centralized or decentralized purchasing

Pain Points & Challenges: Problems your product solves

- Specific challenges or inefficiencies they face
- How urgent these issues are for them

Goals & Objectives: Desired outcomes

- What they're trying to achieve
- How your solution helps reach these goals

Buying Triggers: Events signaling they need a solution

- Funding rounds, leadership changes, growth, regulations, etc.

Success Criteria: How they measure solution value

- KPIs, ROI benchmarks, time saved, etc.
- What "success" looks like after using your product

Ideal Buyer Persona: Who champions the purchase

- Roles typically involved in the buying decision

This is substantial, but you are likely to know some of this from your business planning. Where you have gaps, ChatGPT can help brainstorm potential answers.

Using ChatGPT to draft your ICP

Start with a prompt like this:

Based on the information about my company, generate a clear Ideal Customer Profile (ICP) for our business.

The ICP should include:

```
Firmographics
Technographics
Business Model & Structure
Pain Points & Challenges
Goals & Objectives
Buying Triggers
Success Criteria
Ideal Buyer Persona
Finally, provide a 1-2 paragraph summary of the ideal
customer
```

This sets the framework. ChatGPT uses your company brief to plug in the right details.

How Many ICPs Do You Need?

Three is practical. That's enough variation without overcomplicating things.

Each ICP acts like a compass for your content. Fewer means guesswork. Too many is just noise, so come up with reasonable groupings. Focus on practical, not perfect. You can always refine later.

Review and Refine

When you get a draft, review it carefully. Does it reflect your market reality? If ChatGPT assumes your buyers are larger than they are, clarify: "We target smaller teams, not enterprises." Then re-prompt and refine.

Push Past the First Draft

Don't settle for the initial output, even if it sounds smart. Getting your ICPs just 10% more accurate now pays dividends later. It's compound interest for your marketing.

Ask ChatGPT for ICP variations. Bad examples often inspire good ideas. Compare versions and question the differences. This comparison gives you mix-and-match options to create something better.

Lock It In and Apply It

Once satisfied, save your ICP descriptions. They'll become part of your AI assistant's knowledge base when we configure your custom GPT. You'll reference them often when creating content.

Here are three example ICPs for AnyQuest.ai:

Example Ideal Customer Profiles (ICPs): AnyQuest.ai

ICP 1: Mid-Market SaaS Company (VC-backed, Scaling AI)

This ideal customer is a VC-backed, mid-market SaaS company generating $10M-$100M in annual revenue with 100-500 employees. Located primarily in North America or Western Europe, these companies are in a scale-up phase, often growing rapidly and aiming to optimize operations, customer success, or marketing. They are early adopters of AI and already experimenting with tools like ChatGPT, Zapier, and Slack integrations, but struggle with siloed workflows and inconsistent results.

Their business model is B2B, typically sales-led or hybrid sales/product-led. Buying decisions are made by a centralized innovation or IT leadership team, such as the CIO, Head of Ops, or AI Champion. Pain points include operational inefficiencies in sales research, onboarding, support triage, and knowledge sharing across teams. These companies want to deploy AI workflows faster and more reliably without relying solely on data scientists.

What makes them ready to buy? Signs include stalled internal AI pilots, new AI-focused hires, recent funding rounds, or board pressure to demonstrate ROI from AI initiatives. Their success criteria revolve around speed of deployment, internal adoption across departments, and reduced manual workload. A "win" is an AI agent improving output consistency and reducing repetitive internal tasks.

Champion buyers include the Head of Ops, Director of Innovation, and RevOps leaders who want scalable, compliant, and reusable AI workflows that don't need ongoing developer attention.

ICP 2: Logistics & Supply Chain Operator (Private or PE-owned, Digitizing Ops)

This customer is a privately held or PE-owned logistics or supply chain operator with $50M–$500M in revenue and 200–1,000 employees. Often operating in multiple locations across North America or EMEA, they're in a digital transformation phase, trying to modernize traditionally manual workflows. While not AI-native, they are adopting automation and process improvements rapidly, with moderate tech maturity and a focus on operational efficiency.

The company is B2B, and go-to-market is often partner- or relationship-led. Decision-making is semi-centralized, involving IT and operations leadership. They may use ERPs, document management systems, and communication tools like Microsoft Teams or Outlook. Their pain points include inefficient document review processes (contracts, compliance docs), customer service escalations, and siloed access to internal SOPs or training materials.

They're ready to buy when operational bottlenecks impact service levels, after leadership changes (e.g., new COO or Head of Digital Transformation), or after undergoing compliance audits. Their goals include reducing turnaround time on internal processes, improving visibility across teams, and ensuring consistent policy adherence. A successful outcome is an AI system that integrates with their tools, automates document insights, and routes issues intelligently.

Internal champions might be the Head of Operations, IT Director, or Transformation Lead who are motivated by efficiency, consistency, and data-driven processes.

ICP 3: Professional Services Firm (Mature, Global, Knowledge-Driven)

This ICP represents a mature, often global, professional services firm (legal, consulting, or financial advisory) with $100M–$1B in revenue and 500–5,000 employees. They are typically privately held partnerships or publicly listed, with complex, information-heavy workflows. These firms rely heavily on institutional knowledge and internal content to deliver services. Many have begun adopting LLMs but lack standardized or scalable applications across teams.

Their business model is B2B, and the sales motion is consultative and relationship-driven. Tech maturity varies by department, but innovation leaders and IT teams are keen to centralize AI initiatives. Tools in use often include document management systems (e.g., iManage), internal wikis, Outlook, and increasingly, Microsoft Copilot. Key challenges include time-consuming knowledge discovery, regulatory compliance, and inconsistency in deliverables across teams.

Triggers include adoption of Microsoft Copilot, formation of an AI task force, or rollout of internal AI pilots. They seek faster research capabilities, consistent document analysis, and knowledge reuse across global teams. A win looks like reduced research time, better compliance documentation, and AI tools being adopted firm-wide without adding headcount.

Ideal champions include the CIO, Head of Knowledge Management, and Innovation Directors — all looking for reliable, secure AI agents that understand context and scale across practice groups.

Save and Store the ICP File

It's ok to put all 3 ICPs in the same file. ChatGPT will know what to do. Later, you can reference them by their "Logistics & Supply Chain Operator" or "Professional Services Firm" and it'll understand what ICP characteristics you mean.

In the next chapter, we'll look inside those ideal companies and find who is involved in understanding and purchasing your product. Next up: Identifying Key Buyer Personas.

CHAPTER 7

Mapping Key Buyer Personas

ICPs tell you which companies to target. Buyer personas tell you who inside those companies makes things happen. B2B buying rarely involves just one person – you need to reach decision-makers, users, influencers, and blockers.

Even in B2C, you often deal with different buyer types. Mapping these personas helps your AI assistant create content that connects with the right people at the right time.

> *Tip:*
>
> *Customize the Key Buyer Persona Worksheet in Part 1 of the Appendix for this exercise.*

What is a persona?

A buyer persona is a realistic profile of someone you're trying to reach. Think job role, goals, challenges, motivations, and objections.

Key questions to define personas:

- Job Title and Role: What do they actually do day-to-day?
- Goals and KPIs: What defines success in their role?
- Pain Points: What frustrates them at work?
- Active Problems: What are they currently trying to solve?
- Research Habits: How do they evaluate new solutions?
- Objections: What might make them say no or delay?
- Buying Motivations: What drives their purchase decisions?
- Influencers: Who else affects their decision?
- Communication Style: What messaging resonates – data-driven, practical, visionary?

- Internal Wins: What makes them look good to their boss or team?

Skip surface-level demographics unless truly relevant. Focus on what shapes opinions, how they make decisions, and what drives priorities. This is the deep work AI can't do.

Our Example Personas

We'll focus on three key B2B personas:

Champion

- Internal advocate for your solution
- Motivation: Improve workflow or team success
- Concern: Getting executive approval

Decision Maker

- Final authority on the purchase
- Motivation: Business results, risk reduction
- Concern: Budget, ROI, implementation

Skeptic

- Gatekeeper questioning the need
- Motivation: Stability, avoiding mistakes
- Concern: Security, cost, complexity

Some companies have more roles, but these three provide a strong foundation.

Creating personas with ChatGPT

Now that your AI understands your business and ICP, zoom in on the people involved in buying decisions.

Start with a prompt to identify key persona types:

```
We want to identify key buyer personas involved in
purchasing our solution. Based on our ICP and product,
```

describe the types of buyers who would be involved. For each ICP, give me these 3 roles: who would champion the purchase, who would approve it, and who might be skeptical.

Next, prompt ChatGPT to deepen each role:

Great. Now for each of those personas, create a one-page profile including:

Job Title/Role

Name: a clear, memorable alliterative name

Goals/Motivations

Pain Points/Challenges

Concerns/Objections

Messaging that Resonates

This generates detailed, role-specific profiles to review and refine.

Example Key Buyer Personas: AnyQuest.ai

Buyer Persona 1: The Strategic CIO or CTO

Name: Tech Exec Tom

Role: Chief Information Officer / Chief Technology Officer

Key Motivations:

Deliver scalable AI infrastructure across the organization

Reduce reliance on specialized AI engineers for everyday automation

Align AI efforts with enterprise-wide digital transformation

Pain Points:

Disconnected AI pilots and inconsistent results

Lack of visibility and governance around AI workflows

Executive pressure to show ROI from AI initiatives

What They Care About:

Security, governance, and compliance of AI systems

Seamless integration with existing tools

Centralized control with decentralized execution

Why They Choose AnyQuest:

Visual agent builder reduces engineering bottlenecks

RAG-powered knowledge reuse ensures consistency

Agent-to-agent communication enables scalable orchestration

Success Metrics:

Increased number of AI workflows deployed org-wide

Reduced time-to-deploy AI initiatives

Improved internal adoption and governance of AI tools

Buyer Persona 2: The Head of Operations or RevOps Leader

Name: Ops Olivia

Role: VP of Operations / Revenue Operations / Business Operations

Key Motivations:

Improve team productivity without increasing headcount

Automate repetitive tasks in sales, support, or onboarding

Turn scattered AI usage into repeatable workflows

Pain Points:

High manual effort in recurring tasks

Inconsistent output from AI tools

Low adoption due to technical barriers

What They Care About:

Easy-to-use tools for non-technical teams

Speed of deployment and impact

ROI from operational automation

Why They Choose AnyQuest:

Deploy multi-step AI agents without coding

Templates for common operational use cases

Centralized knowledge used by distributed teams

Success Metrics:

Reduction in hours spent on manual tasks

Improvement in workflow consistency

Adoption rate among operations teams

Buyer Persona 3: The Innovation or Digital Transformation Leader

Name: Innovation Ian

Role: Head of Innovation / Chief Digital Officer / AI Program Lead

Key Motivations:

Stay competitive through cutting-edge AI adoption

Scale successful pilots across departments

Centralize AI experimentation into a sustainable platform

Pain Points:

AI usage confined to technical experts

Difficulty scaling initial wins

Siloed efforts duplicating work

What They Care About:

Fast prototyping and deployment

Low-friction AI adoption

Reusable agents and knowledge capture

Why They Choose AnyQuest:

Enables cross-team reuse of agents

Visual builder for rapid prototyping

Central hub for AI-powered workflows

Success Metrics:

Number of reusable AI agents created

Internal satisfaction with AI tools

Reduction in duplicated efforts

Use real data when you have it. Your goal is 2-4 strong personas to guide content production. Messaging to a technical decision-maker differs greatly from writing to an end-user.

> ### *Tip:*
>
> *Use simple nicknames to make your personas memorable. "Tech Exec Tom" sounds goofy but makes these profiles easier to reference when creating content.*
> *Save these profiles for your custom GPT's knowledge base. Later, when requesting content, specify "this is for Ops Olivia" and the AI will adjust tone and emphasis accordingly.*

Save and Store the Key Buyer Personas File

Same as the ICPs, just put all 3 Buyer Personas in the same file. ChatGPT will know what to do.

Now we have a pretty good idea of the company and the people we must attract and educate. It's time to teach ChatGPT the style and tone it should use to build our content. Next Up: Crafting Your Brand Voice and Writing Style Guide.

CHAPTER 8

Crafting Your Brand Voice and Style

ChatGPT now knows your business and audience. Next, teach it how to speak in your voice. Your brand voice brings personality to your content – without it, even strong messaging feels disconnected.

This step builds Brand Voice Guidelines describing your tone, style, and preferences. Whether friendly, bold, technical, or reserved, documenting these rules ensures AI-generated content reads like it came from your marketing team.

Why brand voice matters (especially for AI)

Your brand voice unifies all content. Without guidance, ChatGPT produces verbose, generic content with telltale AI signs: long sentences, repeated phrases, and unusual text elements like excessive dashes or emojis.

Our goal: content that feels human-crafted with minimal editing. We achieve this by giving explicit style instructions.

> *Tip:*
>
> *Customize the Brand Voice & Style Guide Worksheet in Part 1 of the Appendix for this exercise.*

Defining your voice

Consider how your brand should sound. Are you a trusted advisor? A friendly coach? An innovative expert? If you have existing brand guidelines or strong content examples, keep them handy.

Key elements for voice/style guidelines:

Tone: Confident, conversational, professional, enthusiastic, edgy, warm, humorous, technical, etc.

Personality: Friendly mentor or formal authority? What references or language fits your brand?

Reading level: Aim for upper middle school (clear, straightforward) for broad reach.

Do's and Don'ts: Use short paragraphs and bullet points. Avoid emojis, slang, or excessive hype.

Consistency: Active voice, perspective ("we/our"), and other rules that create a consistent style.

Let's formalize your brand voice using ChatGPT, building on your company brief, ICP, and buyer personas.

Step 1: Analyze and propose a voice

Prompt ChatGPT with:

```
We have information about our company, ideal customer
profile, and key buyer personas. We want to develop Brand
Voice Guidelines for our custom AI assistant.

First, analyze this information and describe what tone,
style, and messaging approach would resonate with our
audience. What should our brand voice be like?
```

```
ChatGPT might respond:
The brand voice should be professional, confident, and
solution oriented. The tone should build trust, avoid
exaggeration, and speak directly to known challenges
and goals.
```

Step 2: Set specific tone and style rules

Next, formalize your voice with this prompt:

```
Draft a Brand Voice & Style Guide based on our company,
audience, and personas.
```

Include sections that cover:

- Audience & Personas

- Tone of Voice

- Writing Style Rules

- Anti-AI Tells

- Content Priorities

- Structure Preferences

Add specific instructions:

- Stick to an upper middle school reading level

- Avoid emojis, em-dash, or generic AI phrasing

- Ask for clarification when unsure

- Sound like a knowledgeable operator, not a marketing machine

Here's a completed example for AnyQuest:

AnyQuest Custom GPT - Brand Voice Guidelines

Audience & Buyer Personas

Our primary audience comprises mid-market to lower-enterprise companies ($1M-$1B revenue) in SaaS, logistics, and professional services. They've begun exploring AI but struggle with secure and consistent scaling.

Key Buyer Personas:

Decision Maker - CIO or VP of Technology

- Focus: ROI, governance, platform alignment

- Needs: Predictable, secure, scalable AI workflows

Champion - Head of Innovation or Digital Transformation

61

- Focus: Faster deployment, cross-team reuse
- Needs: Visibility, measurable impact

Skeptic - CFO, Compliance Lead, or IT Security

- Concerns: Cost, complexity, risk
- Needs: ROI evidence, auditability, integration

Tone of Voice

- Write with clarity, authority, and empathy
- Confident, not cocky
- Clear and direct, not vague or fluffy
- Respectfully persuasive, not salesy
- Future-focused and optimistic, grounded in real value

Writing Style Rules

- Use short to medium paragraphs
- Favor active voice
- Employ bullet points for multiple ideas
- Maintain an upper middle school reading level
- Use plain English and straightforward sentences
- Avoid emojis and em-dashes
- Omit unnecessary transitions (e.g., "moreover")
- Ask clarifying format questions when needed

Anti-AI-Writing Filters

Avoid robotic writing:

- No generic intros ("In today's fast-paced world...")
- Vary sentence structure and paragraph rhythm

- Use specific, practical language—avoid jargon

- Eliminate filler and redundancy

- Skip overly excited claims

- Avoid unnatural phrasing ("synergize your paradigm")

- Every sentence should clarify, address a need, or drive action

Content Prioritization

Prioritize information in this order:

- Reader outcome/value

- Likely current problem

- How AnyQuest helps

- Social proof/use cases

- Clear next step (CTA)

Copy Structure Suggestions

Use scannable, modular blocks with headlines, subheads, and supporting bullets.

Recommended Web Copy Structure:

- Headline: Value or result-focused

- Subhead: Plain language clarity

- Section Body: 2-4 short lines/bullets with real use cases

- CTA: Simple and directive ("Try a workflow")

Email Structure

- Start with a problem

- Highlight the benefit

- End with one CTA

Step 3: Review and Refine the Guidelines

Review your guide carefully. Does it sound like your brand? Shape it to fit. You might add "warmth and a touch of humor are welcome" or remove casual elements for a more formal voice.

Edit the guide yourself or ask ChatGPT to revise:
"This is close, but please include a note about showing empathy and confirm that we always use American spelling."

This guide becomes part of your assistant's core training, ensuring every response reflects your tone and style.

> ### *Tip:*
>
> *ChatGPT has quirks and obsessions. During my course recording, it loved emojis despite instructions to avoid them. Now it's fixated on em-dashes. It'll be something else later.*

This reveals two truths: First, AI makes predictable mistakes, so your editing job remains essential. Second, obvious quirks are easy to spot and fix. The challenge is catching subtle patterns.

If you're repeatedly correcting the same issue, add it to your GPT instructions. But expect imperfection.

With your brand voice established, you now have the three pieces your AI assistant needs:

- Company Brief (your product)
- Personas (your audience)
- Voice Guide (your consistency)

Next, we'll show you how to pull it all together and build your custom AI assistant. Next Up: Setting Up Your Custom ChatGPT Assistant.

PART 4

SETTING UP YOUR CUSTOM
CHATGPT ASSISTANT

CHAPTER 9

Configuring Custom GPTs

I've been there – trying to make AI work for your specific business feels like teaching someone your company language from scratch. Let's fix that by creating an AI that actually knows your stuff.

We'll build a version of ChatGPT that thinks like a member of your team. It remembers your product details, understands your audience, and writes with your voice.

First, I'll walk you through the built-in custom GPT feature (for Plus users). Don't have Plus yet? No problem, I'll show you how to get similar results with a strong prompt in any chat.

Tip:

Customize the Custom GPT Configuration Worksheet in Part 1 of the Appendix for this exercise.

---

1. Create a New Custom GPT

In the ChatGPT interface, log in, then click on your name in the top right. Select the "My GPTs" menu option, then click "Create a GPT." You'll see fields to configure your assistant:

Name: Keep it straightforward. I use names like "AnyQuest Product Marketing Assistant." You might try "[Your Company] Marketing GPT" or something that reflects what it does.

Description: This appears under your GPT's name. Keep it brief: "I help create content for [Company] that aligns with our product and audience."

Instructions: This is where the magic happens. You'll load your training materials – company brief, personas, brand voice – giving your assistant the context it needs to write like it's part of your team.

Here's my version for AnyQuest. Copy this as your baseline and adapt it to your situation:

Role & Purpose

You are a product marketer for AnyQuest.ai.

Your primary goal is to support the marketing and sales efforts for AnyQuest by understanding the product, the market, the personas involved in the sales cycle, and best practices for content creation. You help build marketing materials that are engaging, informative, and conversion-focused.

Knowledge & Context

You have access to the following documents:

- **AnyQuest Minimum Viable Product Brief:** A summary of the company and its products.
- **AnyQuest Persona Document:** Descriptions of key personas in the sales and marketing cycle.
- **AnyQuest Ideal Customer Profiles:** Characteristics of the companies we target.
- **AnyQuest Writing Style Guidelines:** Standards for writing tone, structure, and formatting. Follow these closely.

General behavior rules:

- If unsure, ask clarifying questions before responding.
- Never make up facts or sources.
- Do not use emojis unless explicitly asked.

Tone & Style

Always respond in a tone that is:

- Friendly, witty, and supportive
- Written in short paragraphs using plain language
- Aligned with the AnyQuest Writing Style Guidelines

Workflow & Behavior Rules

Follow this process when responding to tasks:

- Restate or summarize the user's request in your own words.
- Ask a clarifying question if the task is vague or lacks important context.
- If asked to create content and no persona is provided, ask which persona from the AnyQuest Persona Document should be targeted.
- Provide concise and helpful answers that are practical and easy to follow.
- End responses with a suggestion for the next step or a friendly sign-off. If the user agrees the content is finalized, offer to prepare it in a downloadable format.

Optional logic

- If the user says "start over," reset the conversation context.
- If the user says "act like a [X]," adopt that persona until instructed otherwise.

Things to Avoid

- Do not include general disclaimers unless necessary.
- Do not use emojis unless explicitly requested.
- Do not provide legal, medical, or financial advice.

Examples of Good Interactions

- User: Help me write a blog post about AI in customer service.
- You: Got it! You're writing a blog post on AI in customer service. Who's your target audience: Decision Makers, Champions, or Skeptics?

This is where it all comes together. In the instructions field, define the assistant's job, load key context, and set expectations for tone and behavior.

You'll attach files shortly, but for now, copy the essentials: Company Brief highlights, a few lines for each persona, and core style guide rules.

2. Provide Starter Prompts (Optional)

You can add sample prompts that appear as quick-start buttons. These help your team (or future you) see what the assistant does best.

Examples might include:

- "Summarize this product deck into three value props"
- "Write a blog intro for our latest feature release"

I rarely use these myself. Add a couple if they'll spark ideas. Otherwise, you can type your own prompts anytime.

3. Upload Knowledge Files

This step gives your assistant its brain. Check all boxes under *Capabilities* to give full access.

Click *Upload files* and add the documents we've built:

- Brand Voice Guidelines
- Buyer Personas
- Ideal Customer Profiles
- Minimum Viable Company Brief

Make sure each file is enabled by checking the box next to it. This lets your assistant pull from these documents when generating content.

No file-upload option? You probably haven't purchased ChatGPT Plus yet. You should really consider it to get the most value from this course. Without uploads, you can paste content directly into instructions or the chat itself. It works in a pinch, but the assistant won't remember between sessions. Uploading is better.

4. Finalize and Create

Add an image as the finishing touch. This step is optional, but helpful for organization. Upload a logo or use DALL·E (the built-in image tool) to generate something unique.

A custom image might seem minor, but as your collection grows, it makes finding the right assistant easier.

Hit *Save* or *Update*, and congratulations! You've built your custom AI Marketing Assistant loaded with your business context and style rules.

Using Your Assistant

Using your custom assistant feels just like a regular ChatGPT conversation. The difference? It already knows your business. You'll see its name and description on screen.

Test it with a simple prompt:

Give me a one-sentence value prop in a confident tone.

It should return something aligned with your product and voice.

If the output doesn't feel right, maybe it's too vague or includes emojis when it shouldn't, tweak the instructions or recheck the uploaded files. Small changes like "avoid hashtags unless specifically requested" can make a big difference.

Keep the Human in the Loop

Hey look, it's been a couple paragraphs since I reminded you that AI output can be crap. Time for a refresher.

Your assistant is trained, but not perfect. You'll still need to guide and review what it creates. The good news? You don't have to start from zero every time. No more repeating "we're a B2B platform that serves mid-market companies." The assistant already knows that.

In the coming chapters, we'll put this assistant to work on real marketing tasks. We'll generate core messaging, campaign plans, landing pages, emails, and ads. Each time, I'll show you how to prompt effectively and refine the outputs. With your custom AI ready, let's start by developing your core messaging framework.

Next up: Developing Core Messaging Pillars.

PART 5

CORE MESSAGING TO CAMPAIGN LAUNCH

I've seen too many companies where marketing sounds different depending on who wrote it. That's a fast track to confusion. Let's fix that by creating pillars that make consistency easy.

CHAPTER 10

Developing Core Messaging Pillars

Strong messaging is intentional. You need pillars, key points that anchor everything you create. They keep your voice consistent across campaigns, teams, and formats.

In this chapter, we'll use your AI assistant to build those pillars, plus other foundational content like value props, elevator pitches, and taglines.

We'll lean on proven frameworks to guide this process. You've probably heard of some: Jobs to Be Done, StoryBrand, Challenger. Each approach messaging from a different angle. The good news is your assistant can help choose which one fits your product and audience best.

> ### *Tip:*
>
> *While this isn't a "Marketing 101" book, I did put together a bonus chapter on my favorite Messaging Frameworks and when to use/not use each. Check Part 3 of the Appendix for The Hitchhikers Guide to Messaging.*

Selecting a Messaging Framework

Already have a go-to messaging framework? Skip ahead. Though getting a second opinion might help. Your AI assistant can analyze your audience and product, then suggest a framework that matches how your buyers think.

Each framework brings something different:

- **Jobs to Be Done** centers on the task your customer wants to complete.
- **StoryBrand** builds a journey where your product helps the customer succeed.
- **Challenger** messaging focuses on insight that pushes the buyer to rethink assumptions.

Your assistant now has the context to make an informed recommendation, so try this for fun:

Here's a prompt to ask our new assistant to evaluate messaging frameworks:

Analyze the top messaging frameworks used in marketing, including both classic and modern approaches (e.g., Jobs to Be Done, StoryBrand, Challenger, etc.).

Evaluate each framework through the lens of our target personas and ideal customer profiles (ICPs). Then, rank the top 5 frameworks in order of suitability for our needs. For each, provide:

A short description

Why it would be a good fit for our audience and business

Any potential limitations or considerations when applying it

The assistant will give you an analysis like:

- **Jobs to Be Done:** focuses on what the customer is trying to accomplish. Good fit because our personas are goal-oriented; potential downside is it may not highlight emotional appeal.
- **StoryBrand:** frames the company as a guide and customer as a hero. Could work for narrative, but might be too generic for our technical audience without careful application.
- **Challenger:** teaches the customer a new way of thinking, highlighting risks of not changing. Might resonate with our "skeptical" persona by providing insight and urgency. Could be very effective given our solution's newness, but needs strong evidence.

For our walkthrough, let's say Challenger is the right fit. AnyQuest is trying to wake up companies stuck in endless AI pilots, a classic Challenger approach. We're teaching customers what they're missing, then showing them how to fix it.

75

Crafting the Messaging Pillars

> ### *Tip:*
> *Customize the Messaging Pillars Development Worksheet in Part 1 of the Appendix for this exercise.*

Now, armed with our framework, let's prompt the AI to generate core messaging.

Prompt:

Using the Challenger messaging framework, develop core messaging for our company based on what you know about our brand, audience, and offering. Include the following:

A core value proposition

Key messaging pillars (3-5)

Supporting proof points or examples under each pillar

A sample elevator pitch

Suggested headlines or taglines for landing pages or campaigns

Make sure the tone, language, and framing are aligned with our target personas and the problems we help solve.

Example Messaging Pillars: AnyQuest.ai

The previous prompt produces structured output like:

Core Value Proposition: "AnyQuest helps businesses turn scattered AI experiments into scalable, efficient workflows - so they can achieve real results, not just slideshows." *(Notice the Challenger flavor: implying the current state is just "slideshows" while we offer real results.)*

Messaging Pillars:

Operationalize AI for Impact: We take AI out of the lab and into day-to-day operations, closing the "pilot purgatory" gap. *Proof:* "Deploy multi-step AI processes without coding in weeks, not months."

Do More Without More Headcount: Scale AI initiatives without needing to hire large data science teams. *Proof:* "Our platform automates complex tasks; one champion can orchestrate what used to take a whole team."

Integrated & No-Code for Speed: Everything works with your existing tools and doesn't require technical expertise, accelerating adoption. *Proof:* "Plug-and-play integrations and a visual interface mean fast onboarding and results."

Governance & Control Built-In: Safe, secure, and trackable AI deployments to satisfy even the skeptics. *Proof:* "Audit logs, permission controls, and human-in-the-loop options keep you in charge."

Elevator Pitch: "Most companies dabbling in AI are stuck in the sandbox with little to show. AnyQuest changes that. We provide a platform that operationalizes AI at scale, turning isolated experiments into integrated workflows that drive measurable business results – all without needing a team of developers."

Sample Headlines/Taglines:

"Scale Your AI, Not Your Headcount."

"From AI Pilot to ROI."

"Turn AI Ideas into Business Outcomes."

"Operationalize AI: The Fast Lane Out of the Lab."

Take time to review these messages carefully. Do they reflect your product in a way that feels true and compelling? If something's off, adjust it. You can ask the AI to refine specific pillars or generate alternatives with prompts like, "Give me a more direct version of this value prop," or "Rewrite this tagline in a bolder tone."

Keep your personas in mind during review. Every message should connect to a need, concern, or goal that matters to your audience. If you spot a message that doesn't support a key persona, rework or replace it.

For example, if one persona is a skeptical IT manager concerned with risk, you need a pillar addressing governance or security. If your message set only highlights speed and automation, you'll lose their trust.

Once you're satisfied, save this content. These pillars and value propositions become your foundation. Everything you write, landing pages, emails, campaign briefs, builds from here.

Tip:

If it doesn't feel right, dump it and start again. You've barely invested any time at this point.

Refining Messaging

Now that you have main messages, you'll need shorter versions for different formats:

- 10 words: perfect for a tagline or profile bio
- 50 words: works well as a boilerplate or press introduction
- 100 words: ideal for a sales deck or website blurb

Ask the AI to generate versions in each length. Having these ready makes life easier when time is short or space is tight.

I recommend starting with the longer version, then trimming down step by step. The limits might seem arbitrary, but they force you to sharpen your thinking. Once you've created each version, go back and polish them again. Your best phrasing often appears on the second or third pass.

Prompt (optional):

Based on the core messaging we developed, provide succinct company descriptions at various lengths for different uses:

10 words - a tagline or Twitter bio.

20 words - a brief statement.

50 words - a short paragraph for an intro.

100 words - a slightly longer blurb for an "About" section.

250 words - a full paragraph that could be used on a website or in a proposal.

Each version should clearly communicate who we are, what we do, and the value we deliver, while staying aligned with our brand voice and personas.

The AI then produces a set of snippets like:

- 10 words: "Platform turning AI pilot projects into real-world results (no coding)."

- 20 words: "AnyQuest is an AI enablement platform that helps companies scale AI initiatives into everyday operations, no big team needed."

And so on, up to a 250-word narrative recapping your core value and pillars.

Make It So

Work this process until you have a crisp document with great message pillars of various lengths. Ask ChatGPT to format this into a downloadable messaging document in the format of your choice. Then your job is to ensure these messaging pillars are used consistently across your branding, messaging, and collateral. Wasn't that fun?

79

What got us to this point was ensuring our ChatGPT assistant knew about our company, products, and messaging preferences, then we shaped the AI output until it was what we needed. Remember this process, you'll use it again and again to build things with AI.

What's Next

You've now got the message, the voice, and the audience. That's your foundation. Let's put it to work. In the next chapter, we'll build a marketing campaign around this base, planning a concept, outlining a brief, and showing how to turn strategy into something you can actually launch.

CHAPTER 11

Designing an AI-Powered Marketing Campaign

Here's where it gets fun. A campaign connects all these elements to a clear goal, whether that is getting sign-ups, launching a feature, or driving leads.

In this chapter, we'll use your AI assistant to plan a campaign from scratch. We'll choose a theme, draft a campaign brief, and align everything with your core messaging.

Let's say we want to reach the "Innovation Champion" persona. One of our earlier taglines, "Scale Your AI, Not Your Headcount," works perfectly as our campaign theme. Our goal is to generate demo requests from mid-sized companies.

Tip:

Customize the AI-Powered Campaign Planning Worksheet in Part 1 of the Appendix for this exercise.

Step 1: Campaign Strategy Ideation

First, let's ask AI to propose a campaign strategy with a clear goal and persona focus.

Prompt:

Help me design a marketing campaign idea based on this core message: "Scale Your AI, Not Your Headcount."

Goal: Drive qualified leads (Innovation Champions at mid-sized companies) to sign up for a product demo.

Target Persona: "Innovation Champion" (the internal AI advocate at a mid-sized company, eager to implement solutions but constrained by team resources).

Key value prop to convey: Our platform lets you scale AI initiatives without needing to hire lots of new staff.

Suggest:

- A compelling campaign theme or tagline (if different from the core message above),

- The primary channels we should use (e.g., LinkedIn ads, email, webinars, etc. - considering where our persona hangs out),

- and a high-level 4-6 week campaign timeline (major activities by week).

Keep it aligned with our messaging pillars and brand voice.

The AI might respond with something like:

Theme/Tagline: "Scale Your AI, Not Your Headcount" (it may stick with that, or suggest a variation like "Do More with AI, Not More People"). It might also propose a brief narrative for the theme, e.g., focusing on efficiency and smart scaling.

Channels: e.g., LinkedIn Sponsored Content targeting innovation roles, an email drip campaign to leads in our CRM, a live webinar or virtual workshop demonstrating a case study (to engage champions), maybe content marketing like a blog series or whitepaper ("Guide to Scaling AI Without Hiring").

Timeline:

Week 1: Teaser content on LinkedIn and email announcing an upcoming resource or event.

Week 2: Launch a whitepaper or blog post on the theme (gated for lead capture).

Week 3: Webinar event with Q&A (mid-campaign highlight).

Week 4: Follow-up emails to attendees/non-attendees with demo call-to-action.

Week 5-6: Retargeting ads on LinkedIn and Google for those who engaged, with customer success story content.

(This is just an example; the AI might give a slightly different breakdown. The point is a coherent sequence.)

This high-level plan gives us a "bird's-eye view" of our campaign. We have a theme and know what we're doing across channels and time.

Step 2: Writing a Campaign Brief

Now turn your campaign idea into a brief. A brief is a summary document outlining your goal, audience, messaging, and content list. You might share it with your team or use it to stay organized. Let's have the AI format it into something polished and easy to reference.

Prompt:

Now, write a formal campaign brief for the campaign "**Scale Your AI, Not Your Headcount.**"

Include sections for:

Objective: (What are we trying to achieve? e.g., "Book X demo calls from mid-sized enterprise leads in 6 weeks.")

Target Audience: (Who specifically are we targeting? e.g., Innovation leads at mid-sized companies in tech/manufacturing sectors, perhaps mention the persona.)

Key Message: (The core message or value proposition for this campaign.)

Campaign Theme/Concept: (A short description of the creative concept and why it will resonate.)

Channels & Tactics: (List the marketing channels and tactics we'll use, e.g., LinkedIn ads, email sequence, webinar, landing page, etc.)

Timeline: (A brief timeline of campaign phases or key dates.)

Deliverables: (What content pieces or assets will be created - e.g., x ad variations, y emails, landing page, webinar deck, etc.)

Metrics/KPIs: (How we will measure success, e.g., number of demo sign-ups, CTR on ads, etc.)

Format the brief with clear headings and bullet points where appropriate.

Example Campaign Brief: Scale Your AI, Not Your Headcount

The output will be a structured brief like:

Campaign Brief: "Scale Your AI, Not Your Headcount"

Objective: Generate 50 qualified demo requests from mid-sized enterprise prospects over a 6-week campaign.

Target Audience: Primary focus on "Innovation Champions" - mid-level to senior ops/innovation managers at companies (~$50M-$1B revenue) who are experimenting with AI solutions. Secondary audience includes their leaders (CTOs, COOs who would approve the purchase).

Key Message: You can achieve large-scale AI implementation without a large team. Our solution allows you to leverage AI across your organization efficiently, so you get results while saving on hiring and onboarding.

Theme/Concept: *"Scale Your AI, Not Your Headcount"* - The campaign plays on the fear that adopting new tech often means needing more people. We challenge that

84

notion, showing how smart tools eliminate the need for extra headcount. Creative visuals might feature a rising graph labeled "AI output" versus a flat line "team size," illustrating doing more with the same team. Tone will be optimistic and empowering, appealing to the champion's resourcefulness.

Channels & Tactics:

LinkedIn Sponsored Content: Targeting job titles like "Innovation Manager, Operations Director, Digital Transformation Lead" with carousel ads and single-image ads focusing on the theme.

Email Sequence: 3-part email series to our prospect list:

(1) Introduction to the concept with a success story teaser,

(2) Invitation to a webinar or to download a guide,

(3) Final call encouraging demo booking by highlighting scarcity or urgency ("Don't get left behind while others scale up").

Webinar/Virtual Event: A 30-minute live demo + Q&A in Week 3 showcasing how our platform works without coding, featuring a case study. This engages champions and allows them to invite their boss.

Landing Page: Dedicated page "Scale AI" with campaign messaging, where all traffic from ads/emails is directed. Contains benefits overview, brief case study, and demo request form.

Content Marketing: A downloadable guide titled "5 Ways to Scale AI Initiatives Without Hiring" gated on the landing page to capture leads. Promoted via LinkedIn and email.

Retargeting Ads: Google Display and LinkedIn text ads reminding visitors who didn't sign up to take action.

Timeline:

Week 0 (Prep): Create assets (landing page live, ads creative, email copy, webinar content ready).

Week 1: Launch LinkedIn ads and send Email #1.

Week 2: Continue ads, send Email #2 mid-week announcing upcoming webinar.

Week 3: Host Webinar (mid-week). Send thank-you email with recording + CTA to attendees, and "sorry we missed you" with recording + CTA for no-shows.

Week 4: Send Email #3 (final call to action, e.g., "Last chance to see how to scale AI easily").

Week 5-6: Continue follow-up with personal outreach from sales on hot leads, run retargeting ads. Wrap up and measure results at end of Week 6.

Deliverables:

Ad creatives: 2 LinkedIn single image ads, 1 LinkedIn carousel (with 3 cards), and 2-3 retargeting ad variants.

Emails: 3 marketing emails (html design + text).

Webinar: Slide deck + demo script, and recording to use as content.

Landing Page: Copy and design for the campaign page, plus form integration to CRM.

PDF Guide: "Scaling AI Without More People" (5-6 page PDF) for download.

Metrics/KPIs: Track number of demo requests (primary KPI). Also monitor LinkedIn ad CTR (aim >0.5%), landing page conversion rate (aim 15%), webinar sign-ups and attendees, and email engagement (open/click rates). Post-campaign, track how many demos convert to opportunities for longer-term ROI.

> ### *Tip:*
>
> *As you build assets you like, hold on to them. Feed them to ChatGPT as "Good" examples and it'll (mostly) follow that structure.*

Now that your campaign is mapped out, let's create content. The landing page is typically the first and most important piece to educate and encourage the next steps to action. In the next chapter, we'll use the AI to create a landing page outline and write copy that matches your messaging and tone. Next Up: Creating High-Converting Landing Pages.

CHAPTER 12

Creating High-Converting Landing Pages

Your landing page is the campaign centerpiece. It's where your ads and emails drive traffic, and where you convert interest into action (sign-ups, inquiries, purchases). For our "Scale Your AI" campaign, we need a dedicated page that continues the story and convinces visitors to request a demo or download our guide.

In this chapter, we'll generate a landing page structure and copy using our AI assistant. We'll focus on the elements of persuasive content: strong headline, engaging hero section, benefit bullets, social proof, and clear call-to-action (CTA).

Step 1: Outline the Landing Page Sections

Let's have the AI outline the page sections first. A typical B2B landing page includes:

- **Hero section:** Headline, subheadline, and CTA at the top.
- **Problem/Pain section:** Identifying the pain point.
- **Solution/Value section:** How your product addresses the pain.
- **Benefits or Features bullets:** Key benefits with short descriptions.
- **Social Proof:** Testimonials, customer logos, or results.
- **CTA section:** Another call-to-action near the bottom.
- Possibly an FAQ or objection-handling section for longer pages.

Let's prompt the AI for a recommended structure:

Prompt:

Outline a landing page for the "**Scale Your AI, Not Your Headcount**" campaign, targeting our Innovation Champion persona.

The landing page should include:

A compelling **headline** and subheadline that grab attention (address the pain of feeling stuck or understaffed, and promise a solution).

A brief **overview paragraph or video pitch** (we'll focus on text now, but it should work as a quick intro).

Key benefit points (3 bullet points) highlighting what they gain (e.g., scale AI projects faster, no big team needed, proven results).

A prominent **Call-to-Action (CTA)** (like a button "Book a Demo" or "Get the Guide") near the top.

A **social proof section** - perhaps a testimonial quote or a statistic (you can make up a realistic stat like "Used by X companies to automate Y tasks").

Optionally, a **visual or diagram suggestion** that could accompany (like an illustration of doing more with same team).

A final **CTA section** reinforcing why they should act now.

Just list the sections and a brief description of what goes in each, not the full copy yet.

Example Landing Page Structure: Scale Your AI, Not Your Headcount

The previous prompt produces a structure like:

Hero Section: Headline + subheadline + CTA. (Headline idea: "Do Big Things with AI – Without a Big Team." Subhead: "Discover how to operationalize AI across

89

your company without hiring an army of developers."
CTA: [Button] "Request a Demo").

Intro/Overview: A short paragraph acknowledging their challenge (lots of AI ideas, limited team) and introducing our solution as the answer.

3 Key Benefits:

"Multiply Your Impact" - Achieve more AI-driven outcomes with your existing team.

"Fast Implementation" - Deploy AI workflows in weeks, not months, no coding required.

"Proven Results" - See immediate ROI and success stories from companies like [Customer X].

Social Proof: A quote from a happy customer or a statistic: e.g., *"Thanks to [Our Product], we automated 50+ processes and saved 1,000 hours - without hiring extra staff,"* - **Jane Doe, Operations Director**. And/or logos of 3-5 companies using the product.

Closing/CTA: A final pitch line like "Ready to scale your AI initiatives efficiently?" and a CTA button or form again ("Schedule My Demo" or "Download the Guide").

Now we have a blueprint of sections to fill in.

Step 2: Generate the Landing Page Copy

Next, let's have the AI write the actual copy for each section.

Prompt:

Write complete landing page copy for AnyQuest's campaign: "Scale Your AI, Not Your Headcount."

Details:

Persona: Innovation Champion at a mid-sized company experimenting with AI

Core message: AnyQuest helps you build, scale, and operationalize AI workflows across teams—without technical friction

CTA: "Book Your Demo"

Include:

Headline (H1) and Subheadline (H2) for the hero.

A short body paragraph for the intro.

3 bullet point benefits (with short explanations, one sentence each).

A placeholder for a testimonial (you can fabricate a realistic one).

And a strong closing statement/CTA inviting the user to take action (like scheduling a demo).

Ensure the copy is in our brand voice (confident, insightful, to-the-point) and speaks to the Innovation Champion persona's motivations and concerns.

Keep it persuasive but not overly salesy. Use the second person ("you") where appropriate to engage the reader.

Example Landing Page Copy: Scale Your AI, Not Your Headcount

The assistant will give us text for each part:

Headline: Scale Your AI, Not Your Headcount.

Subheadline: Unlock enterprise-level AI results with the team you have – no extra hires needed.

Intro Paragraph: Are your AI pilots stuck in limbo because you don't have a dedicated data science army? You're not alone. Many innovation leads like you have ambitious AI goals but limited resources. Our solution

91

changes the game: it allows you to deploy powerful AI workflows across your organization quickly, without hiring a single new engineer. We help you transform what you already have – your team and your data – into tangible business results.

Key Benefits:

Do More with Less: Amplify your team's impact. Our platform enables one champion to accomplish the work of a full AI squad, through automation and smart workflows.

Fast, Code-Free Deployment: Launch AI initiatives in weeks, not months. With our no-code interface, you bypass IT bottlenecks and start delivering value immediately.

Proven at Scale: Get results you can brag about. (In fact, one client automated 50+ processes and saved 1,000 hours in a quarter.) We bring success stories and expertise so you can confidently lead the AI charge.

Testimonial (Social Proof): "Using [Product], we turned an AI 'science project' into real operational workflow in just 30 days. And we did it without any new hires. It's like I suddenly had an expert AI team at my fingertips."
— **Jane Doe, Director of Operations, TechCorp Inc.**

(Or a simple line: Trusted by teams at [Client A], [Client B], and [Client C].)

Closing & CTA: You don't need a big team to achieve big things with AI. **AnyQuest** equips you to scale initiatives now, on your terms. Don't let limited headcount limit your vision.

Ready to see what you can do?

[Request a Demo] and discover how to turn your AI ideas into business breakthroughs.

That's a solid first draft! It addresses the pain point (stuck AI pilots), offers the promise (do it without new hires), provides specific benefits aligned with our messaging pillars, and includes a testimonial for credibility.

Review and Polish Checklist:

- **Headline:** Grabs attention and reflects the campaign name.
- **Tone:** Confident and direct, using "you" to speak to the reader.
- **Benefits:** Relevant to the persona's priorities: do more with less, fast deployment, and real outcomes.
- **Testimonial:** Solid placeholder. Replace with a real quote when available.
- **CTA:** Clear, action-oriented, and easy to understand.

The structure and content are strong. For a real page, you'd want to check character counts for headlines and possibly add a supporting stat, but this is ready to go.

The landing page copy is now finished. You'd typically plug this into your site or landing page tool, but from a messaging standpoint, we're set.

In the next few chapters, we'll focus on driving traffic and following up by creating an email sequence then ad copy designed to attract the right audience on platforms like Google and LinkedIn.

Next up: Writing Effective Email Outreach Sequences.

CHAPTER 13

Writing Effective Email Outreach Sequences

Email still works. Not like it used to, but when done right, it moves people from curious to committed.

Here's the thing most people get wrong: they treat email like a megaphone instead of a conversation. Your prospects aren't waiting around for your next brilliant message. They're drowning in their inbox, just like you.

In this chapter, we'll focus on follow-up sequences for people who showed interest but haven't taken the next step yet. Think of someone who downloaded your guide or signed up for a webinar but hasn't requested a demo. They're warm, but not quite ready to buy.

We're going to craft a three-email sequence that gently nudges them forward without being pushy. Here's how it works.

Email 1: The Thank You (With a Soft Nudge)

This email triggers right after someone engages with your content. Your goal? Thank them and invite them to take the next logical step.

Most people rush straight to the sales pitch. Don't do that. Start with gratitude, then acknowledge their situation before making your ask.

Try this prompt for your AI assistant:

Draft an email to someone who just engaged with our "Scale Your AI, Not Your Headcount" campaign (for example, they downloaded our guide or attended our webinar).

Purpose: Thank them for their interest and invite them to take the next step (book a demo).

Audience: Innovation Champion persona (mid-level manager interested in implementing AI).

Tone: Friendly, helpful, and consultative - we're positioning ourselves as a partner.

Length: ~100 words.

Include:

- A thank you for what they did (e.g., "Thanks for downloading our guide!")

- A sentence or two that connects their interest to a potential need ("Companies like yours often struggle with X... hope the guide gave you ideas")

- A clear call-to-action: inviting them to a demo or call, highlighting the benefit ("see it in action" or "tailored consultation").

- Sign off from a real person (e.g., you or a team member name/title) to feel personal.

Example Email 1: Scale Your AI, Not Your Headcount

Your AI might generate something like:

Hello [Name],

Thank you for downloading our "5 Ways to Scale AI Without Hiring" guide! I hope you found some useful strategies in there for expanding your AI projects with your current team.

Many companies I speak with are in a similar spot. They have big goals, but limited resources. The good news: there are ways to get AI wins without adding headcount (as you saw in the guide).

If you're curious to see how this could work specifically for your organization, I'd love to offer a quick demo of AnyQuest. In 30 minutes, we can show you how to automate some of the AI tasks on your plate and answer any questions you have.

95

Interested in seeing it in action? Just reply to this email or click the link below to pick a time for a demo.

Cheers,

[Your Name]

[Title], [Company]

Notice what this does well: it's personal, acknowledges their challenge, and frames the demo as helpful rather than sales-y. The call-to-action is simple and gives them options.

Email 2: The Gentle Reminder (With Added Value)

If they don't respond after a few days, send a gentle follow-up. Don't just repeat yourself. Add something new that shows you have more to offer.

Prompt for Email 2:

Now write a follow-up email for those who didn't respond to the first email.

Goal: Remind them of our value and prompt action, without nagging.

Tone: Warm, professional, no pressure.

Content:

- Reference the previous email/webinar briefly ("I know things get busy...")

- Reiterate a key benefit or an insight they might have missed ("By the way, one idea from the guide was X, which could potentially save you Y...")

- Again invite them to chat or demo, maybe offering to answer any specific question they have.

- Keep it short (~100 words).

Example Email 2: Scale Your AI, Not Your Headcount

Here's what you might get:

```
Hello [Name],

I know things get busy, so I wanted to follow up in
case my last email got buried. We recently shared some
tips on scaling AI without extra hiring - one idea was
to streamline repetitive tasks with AI agents (which
could save teams dozens of hours a month).

If challenges like that are on your mind, we'd be happy
to discuss solutions. No pressure - even a short chat
might spark an idea or two that you can use.

Would you be open to a quick call or demo? Feel free
to reply with any question or click below to schedule.

Thanks,
[Your Name], [Company]
```

This email works because it doesn't whine about being ignored. Instead, it offers genuine insight while keeping the door open.

Email 3: The Graceful Exit (That Sometimes Works Magic)

If there's still no response, it's time for a "breakup email." This might sound counterintuitive, but telling someone you'll stop emailing them often prompts a response.

Why? Because it removes pressure and shows respect for their time. Sometimes people are genuinely interested but just overwhelmed. Also, they may be putting off responding until later, and having a concrete reminder your time is also valuable can nudge them into action.

Prompt for Email 3:

```
Lastly, write a brief "breakup" email to a lead who hasn't
responded to previous two emails.
```

Tone: Casual, respectful, open-ended.

Content:

- A subject line that signals this might be the last email (e.g., "Should I close your file?" or something light).

- Express that we don't want to bother them and this will be the last reach out.

- Invite them one more time if they're still interested, or to let us know if now isn't a good time.

- Maybe even a gentle humor or acknowledgement ("I promise this is the last you'll hear from me unless you say otherwise").

- End on a friendly note, keeping the door open.

Length: ~75 words.

Example Email 3: Scale Your AI, Not Your Headcount

Your AI might create:

Hi [Name],

Guessing your plate is pretty full (I totally get it). I didn't want to keep filling up your inbox, so this will be my last email unless I hear back.

If scaling AI is still on your radar, I'm here and happy to chat on your timeline. And if now isn't the right time, no worries at all.

Either way, thanks for your time and have a great day!

Best, [Your Name]

Here's what makes this work: it's genuinely respectful, not manipulative. You're actually backing off, which paradoxically makes people more likely to respond.

Subject Lines Are Everything

Your subject line is make-or-break. It's the bouncer at the door of someone's attention, and most of the time, it's saying "not today."

It's important to remember the mindset of someone scanning their inbox is usually "What can I delete?". Your job is to catch their attention before you get sent to the trash bin.

Think about your own inbox. You scan subject lines in milliseconds, deciding what deserves your time. Your prospects do the same thing. The difference between "opened" and "deleted" often comes down to those few words at the top.

A good subject line does three things: it's clear about what's inside, it creates just enough curiosity to earn the click, and it feels like it comes from a real person. When you're following up after someone downloads your guide or attends your webinar, that subject line becomes the bridge between their initial interest and acting.

Here's what works.

10 Subject Line Rules That Get Opens

1. **Keep it short.** Under 50 characters means it won't get cut off on mobile. Most people check email on their phone first.
2. **Sound human.** Write like you're talking to a colleague, not broadcasting to the masses. "Quick question" beats "Maximizing Your Marketing ROI" every time. (re-read this. Marketers default to 1 to many language but email is 1:1)
3. **Be specific.** "Checking in" is marketing speak for "I have nothing valuable to say." Try "The #1 mistake I see after people download this guide" instead.
4. **Use curiosity, not clickbait.** Create genuine intrigue without being misleading. You want them curious, not annoyed.
5. **Make it timely.** Reference something recent: their download, a conversation, or what's happening in their industry right now.

6. **Personalize when it matters.** Use their name or reference what they actually did. "Thanks for downloading the AI guide, Sarah" feels more personal than "Thanks for your interest."

7. **Try numbers.** "3 ways to use AI in your next campaign" often outperforms vague statements. Numbers feel concrete and actionable.

8. **Ask a question.** Questions naturally trigger engagement. "Still considering AI for Q3?" works because it assumes ongoing interest.

9. **Match your tone to their journey.** Early in the relationship? Stay friendly and helpful. Later in the process? You can be more direct.

10. **Break the pattern.** "Should I close this file?" works precisely because it doesn't sound like every other follow-up email.

Write to the Moment, Not the Metric

The best subject lines pick up where you left off. If someone just downloaded your guide, your next email should feel like the natural next chapter: "Here's what most people do next."

If they opened your last email but didn't respond, try something like "Still on your radar?" It acknowledges that they saw your message but didn't act on it.

Your job isn't to be clever. It's to be useful and real. Write like someone who actually remembers the conversation you just had. Because when your email feels like part of a relationship instead of part of a campaign, getting them to open is just the beginning.

Try this prompt with your AI:

You're a marketing copywriter helping me improve the subject line of a follow-up email. The context is: [briefly describe what the email is about - e.g., "Following up after someone downloaded our AI Scaling Guide"].

Generate 5 subject line variations, each using a different approach:

```
Direct and clear

Curiosity-driven

Conversational and casual

Urgent or time-sensitive

Humorous or unexpected

Keep each subject line under 50 characters. Make sure they
relate to the email content but stand out in a crowded
inbox.
```

Your AI will give you options that feel human and relevant to exactly where your prospect is on their journey. Remove the worst 3, then A/B test the other two.

> ### Tip:
>
> *Always ask for several versions with specific tones.*

Character Limits Are Real (And AI Gets Them Wrong)

Large language models think in words and concepts, not individual letters. They'll get you close, but you'll need to trim and adjust.

This is one of those "human in the loop" moments. Let the AI do the heavy lifting, then step in to polish. We'll cover this in more detail in the next chapter.

Ready to dig into writing ads? Next Up: Beyond Email: Ad Copy That Actually Works.

PART 6

BEYOND EMAIL: AD COPY THAT ACTUALLY WORKS

Your email sequence is just part of the funnel. Before you can nurture leads, you need to attract them. That's where ad copy comes in.

Whether it's Google or LinkedIn, you've got seconds to convince someone to click. Your copy needs to be clear, tight, and focused on what matters to them.

CHAPTER 14

Google Search Ads: Short and Structured

Google ads are all about matching intent. Someone types "AI automation platform" and your ad needs to say "Yes, this is exactly what you're looking for."

You get several headlines (30 characters each), two description lines (90 characters each), and a display URL. Use them wisely. Every single word and character has to fight for its place.

Unlike your blog posts or social content, these ads are being judged in milliseconds by people who are already hunting for what you offer. They typed "AI automation platform" into Google because they need exactly that. Your job is to prove you're the answer to their search.

If your copy mirrors their intent and speaks their language, you've got a shot. If it's vague or generic, they'll scroll past without a second thought. The good news? AI excels at helping you test different angles until you find what clicks.

"Constraints breed creativity."

One of my business mentors used to say "Constraints breed creativity." He was right, but he forgot to mention they also breed a special kind of insanity. I've spent way too many hours staring at a screen, trying to cram one more idea into 29 characters.

Here's where AI becomes your best friend. Instead of wrestling with one version for hours, you can generate dozens of variations and pick the best ones. Let the machine do the heavy lifting while you focus on strategy.

Try this prompt:

Write Google Search ad copy for the following campaign:

Product/Service: AnyQuest AI Automation Platform

Target audience: Innovation Champions at mid-sized companies

Campaign theme: Scale Your AI, Not Your Headcount
Goal: Drive demo requests from qualified leads
Include:
3-5 headline options (max 30 characters each)
2 description options (max 90 characters each)
1 call-to-action variation
Tone: Clear, confident, and benefit-driven

Example Google Ad Copy: Scale Your AI, Not Your Headcount

Your AI might generate:

Headline options (<=30 chars):

Scale AI Without More Staff

Do More AI, No New Hires

AI Results, Same Team

Operationalize AI Fast

AI Automation Platform

Description options (<=90 chars):

Scale your AI projects in weeks, not months – with your current team. No coding or extra hires needed.

AI solution that multiplies your team's output. Achieve more (without adding headcount). Get a free demo.

The key is mixing benefit-focused headlines with outcome-focused ones. When Google rotates them, they tell a complete story.

It's often a bonus to include the call to action in a headline, like "Get a Demo." But you need to watch the character limits when combining lines.

In our case, we already have "Get a free demo" in the description, so we're covered.

Now, about those character counts...

Reality Check: AI Math Doesn't Always Work

We should double-check character counts:

1. "Scale AI Without More Staff" (27 chars) – good.
2. "Do More AI, No New Hires" (24 chars).
3. "AI Results, Same Team" (21).
4. "Operationalize AI Fast" (22).
5. "AI Automation Platform" (22).

All under 30. Nice job ChatGPT. Now let's examine the descriptions:

- Scale your AI projects in weeks, not months – with your current team. No coding or extra hires needed. (102 chars) ??
- AI solution that multiplies your team's output. Achieve more (without adding headcount). Get a free demo. (105 chars) :(

I asked the AI to keep descriptions under 90 characters. It gave me some that were 105 characters. This happens all the time.

AI thinks in concepts, not character counts. It'll get you in the ballpark, but you'll need to trim the fat yourself. Don't get frustrated. Just factor in some editing time.

Pardon My Rant

I've been hammering this point throughout the book: you can't just trust AI output and walk away.

Your AI gives you a 35-character headline when you asked for 30? That's obvious. You can count to 35. Gold star for you.

But figuring out if your ebook section is rambling for three paragraphs with no real point? That's harder to spot, and it's where AI often goes off the rails.

106

> ### *Tip:*
> *Quantitative problems jump out at you. Qualitative problems hide in plain sight.*

When AI writes your sales page, it might technically check every box on your list while somehow reading like it was written by a very polite robot who learned English from LinkedIn posts (Because it did).

When it crafts your email sequence, the tone might be just off enough to make you sound like every other "growth-focused, results-driven, customer-centric" company that ever existed.

Your prospects won't think "this headline is two characters too long." They will think "this doesn't feel right for me" and move on. And *you* might not catch it because the content looks professional and covers all the points you asked for. Plus, who has time to read all those words?

You own the outcome. When that email campaign flops or that landing page converts poorly, it's not the AI's fault. It's yours for not catching what the machine missed.

Try this: After AI generates your content, read it out loud (or get a text-to-voice to do it). Does it sound like something you'd actually say to your customer? If not, that's your qualitative red flag. Fix it before it goes live.

There's more to life than Google ads, especially for B2B products and services. LinkedIn offers richer firmographic and business persona targeting than anywhere else. Next up: we build LinkedIn Sponsored Content.

CHAPTER 15

LinkedIn Sponsored Content: Conversational and Professional

LinkedIn ads are a completely different beast than Google Search ads. With Google, someone just typed "help me find a solution to this specific problem I have right now." They're motivated, focused, and ready to buy something.

On LinkedIn, people are scrolling through their feed while pretending to work. They're catching up on industry gossip, humble-bragging about their latest promotion, or avoiding that email from their boss. Your ad needs to politely tap them on the shoulder and say "Hey, this might actually matter to you."

You're not matching keywords anymore. You're earning attention from people who didn't ask for it. That changes everything about how you write.

What Works When Attention Spans Don't

One of the strengths of LinkedIn ads is laser precision targeting. Want to reach directors of marketing at mid-sized SaaS firms who went to state schools and drink oat milk lattes? LinkedIn's got you covered. (Okay, maybe not the latte part, but you get the idea.)

Targeting is one part of the puzzle, but mindset is the other. Your ad viewers are scrolling between updates about someone's work anniversary and another thought leader's hot take on remote work. They're distracted, mildly curious, perhaps professionally ambitious, but definitely not in buying mode.

Your copy needs to work twice as hard. You need to resonate with "Show Me You Know Me". Start with a pain point that makes them go "Oh, that's me" or hit them with an insight that stops their scroll. Ask a question that makes them think or tease a takeaway that makes them curious.

Keep that intro tight. Your goal is to earn the click before they hit "See More" and move on to someone's vacation photos.

LinkedIn Sponsored Content Ad

For LinkedIn, a sponsored content post usually has:

- An intro text (often a sentence or two that shows above the image).
- A headline (appears below the image, somewhat like a title, shorter).
- Possibly a description below that (depending on format).
- But LinkedIn's main one is usually just the post text and maybe a short headline if it's a link ad.

Try this prompt:

```
Write LinkedIn ad copy for a Sponsored Content post.
Details:
Product/Service: AnyQuest AI Automation Platform
Audience: Innovation Champions at mid-sized companies
Goal: Book a product demo
Message: Scale AI Without Scaling Your Headcount
Generate:
1 attention-grabbing headline (max 70 characters)
1 short post (under 150 words)
1 CTA (e.g., Learn More, Get the Report, Book Demo)
Tone: Professional but conversational
```

Example LinkedIn Ad Copy: Scale Your AI, Not Your Headcount

You might get:

Headline: "Scale AI Without Adding Headcount"

Post Text: Still stuck in AI pilot mode? You're not alone.

```
Most mid-sized companies have the ambition but not the
bandwidth to fully scale their AI initiatives. Sound
familiar?

AnyQuest makes it possible to automate complex
workflows without hiring a team of engineers. Our
visual builder and built-in templates help you move
from pilot to production without the overhead.

If your AI efforts feel stalled or siloed, it's time
to unlock progress without increasing your team size.

See how forward-thinking teams are scaling AI
efficiently.

Call to Action: Book Demo
```

This LinkedIn ad hits all the right notes. It opens with a problem that makes people nod their heads, offers a clear path forward, drops in some social proof without being pushy, and ends with an obvious next step. The copy feels conversational and targeted, like it was written specifically for the person reading it.

LinkedIn rewards brevity, and this ad delivers. It sits right at that sweet spot just above the "see more" cutoff where most people decide whether to keep reading. You could trim a word or two if you're feeling fancy, but honestly? It works as-is.

It's all fun and games until it hits the real world. Our first efforts won't be the best and we need the market to vote so we can refine.

Testing Different Angles (The Easy Way)

Want to test different approaches? Just ask your AI for variations:

- "Give me one headline about saving time and one about saving money"
- "Write two opening lines, one friendly and one more urgent"

110

Boom. You've got multiple angles to test in about 30 seconds. Try the time-saving angle with one audience segment and the cost-saving angle with another. See which one gets more clicks, then double down on the winner.

Just remember our old friend the character count issue. Even if ChatGPT generates brilliant ideas, you'll still need to double-check the lengths when you upload them to the ad platform. AI is creative, not mathematical.

PART 7

EXPAND BEYOND CAMPAIGNS

Step back for a second and look at what you just created:

- Ads that attract the right people
- Landing pages that inform and convert
- Email sequences that nurture without being annoying
- An AI assistant that made all of this faster and more consistent

You just created a complete and connected funnel system. Each piece connects to the next, moving people from curious to committed in a way that feels natural, not pushy.

This is just the beginning.

How do you spot what's working? How do you improve your AI's performance? And how do you scale this across your entire marketing operation?

More magic happens when you start iterating and improving based on real results. Next Up: Strategic Marketing Tasks

CHAPTER 16

Strategic Marketing Tasks

Let the Machine Do the Grunt Work

I've watched countless teams get stuck in tactical weeds, using ChatGPT like a fancy grammar checker. Meanwhile, the real opportunity sits right there, waiting. Strategic work. The stuff that moves the needle.

Let me tell you what I mean.

A client called me last month, frustrated. Their product launch had flopped, and they couldn't figure out why. Marketing blamed the messaging. Sales blamed the leads. Customer success blamed the onboarding. Sound familiar?

They had feedback scattered across Slack threads, meeting notes, and random email chains. Pure chaos. Buried in that feedback was gold. They just needed someone to pan for it.

We fed everything into ChatGPT and asked for a clean retrospective. It grouped themes, organized by funnel stage, then made specific suggestions for round two. The whole team finally saw the same picture.

That's strategic AI. You're not outsourcing judgment. You're accelerating insights and enablement.

Campaign Post-Mortems That Actually Help

Most post-mortems are theater. Everyone sits in a room, shares opinions, and nothing changes. But if you've captured feedback during your campaign, you can skip the drama.

Here's how it works: collect everything. Meeting notes, survey responses, sales objections, support tickets. Then ask ChatGPT to find the patterns.

Group feedback by channel. Organize insights by buyer stage. Surface the themes that matter.

You'll get something you can use instead of another meeting that could have been an email.

Try this prompt:

You're a senior B2B marketing strategist.

Use this product launch feedback from marketing, sales, and CS to generate a campaign retrospective:

3 things that went well, 3 that didn't, and 3 action steps for next time.

Market Research That Doesn't Take Forever

Competitive analysis used to mean days of browsing websites and taking screenshots. Now? Copy your top five competitors' homepage messaging, plus what you can grab from Capterra, paste it into ChatGPT, and ask what you're missing.

You'll spot feature gaps you hadn't noticed. Positioning angles you could own. Messaging that everyone's using (which means you should probably avoid it).

The same approach works for customer journey mapping. Describe your ideal buyer, their pain points, and their decision process. ChatGPT will draft journey stages and suggest messaging for each phase. It's not perfect, but it beats staring at a blank whiteboard for three hours.

Try this approach: Feed in competitor copy and ask:

What messaging themes do you see?

What's everyone saying?

What angles are missing?

115

What I want you to feel at this point in the book is not how much less you have to work, but how much better and more complete of a job you can do with AI at your side. With that in mind, we now move into how you can get to the stuff you've meant to work on, but just haven't had the space or time in your day. Next Up: Creative Content Beyond the Basics.

CHAPTER 17

Creative Content Beyond the Basics

Let's talk about creative content. The stuff that makes people feel human on the other end of your marketing.

Here's what I've learned after years of watching teams struggle with content creation: the blank page problem isn't really about creativity. It's about starting.

You know how it goes. You need podcast questions for next week's guest. Or a webinar outline for that feature launch. Or interview questions that don't sound like they came from a marketing template. You sit there, cursor blinking, waiting for inspiration.

Stop waiting.

If you've got raw material, you can transform it. Guest bio and LinkedIn profile? That's enough for a solid question list. Customer interview transcript? That's a case study waiting to happen. Internal product notes? That's your webinar outline.

The trick isn't having ChatGPT write your content. It's having it help you shape what's already there.

Podcast Planning Without the Panic

I've been there. Guest confirmed, recording scheduled, and you're scrambling for questions the night before. Here's what works better:

Grab their bio, website copy, and any recent content they've shared. Feed it to ChatGPT with context about your audience. You'll get a starter set of questions connected to their expertise.

Are they perfect? Probably not. But they're specific, relevant, and way better than "So, tell us about your background."

Try this prompt: "Take this guest bio and website copy and generate 8 interview questions for a B2B marketing podcast. Focus on tactical insights our audience can use next week."

Customer Stories That Actually Sell

Here's something most people miss about customer stories: the transcript is usually better than the final case study.

Real conversations have energy. They have personality. They have the messy, human moments that make stories believable. But then we sanitize them into corporate speak and wonder why no one reads them.

Instead, take that raw transcript and ask ChatGPT to structure it, not rewrite it. Keep the voice. Keep the specific details. Just organize it into something people can scan and use.

Try this prompt:

```
Take this customer interview transcript and turn it into a
300-word customer story with four parts:
Problem

Solution

Outcome

One specific detail that makes it memorable.

Keep their authentic voice.
```

Endless Possibilities

Step back from the specifics for a moment. See how we take known good content and transform it into something else? This is the pattern you will use to push your efforts forward at a speed you didn't know was possible.

I spent many years as a product engineer, building software products. The skills I used were in breaking down a business need into the data, rules, and visual components needed to complete business tasks. I think of content projects the same way.

118

- **Data**: Your source material.
- **Rules**: Prompts and context needed to transform the material
- **Visual Components**: What output do I need?

For marketers, just like you wouldn't redesign your logo for every campaign, you shouldn't reinvent your content process for every piece. Think of AI content creation like the process of developing a signature dish for your restaurant. You'd build on proven ingredients and techniques, then adapt them for your specific audience. The AI is your sous chef, helping you combine these elements into something that serves your audience.

A chef breaks down a complex dish into prepped ingredients, cooking techniques, and plating. You break down content projects into source materials, transformation prompts, and desired formats.

The Content Kitchen Process:

1. **Gather your ingredients** (source materials: interviews, data, existing content)
2. **Choose your cooking method** (prompts and context for transformation)
3. **Prep your mise en place** (organize materials by content type and audience)
4. **Cook with technique** (apply AI prompts systematically)
5. **Plate for presentation** (format for your specific channel and audience)
6. **Taste and adjust** (review, refine, make it authentically yours)

The AI helps you remix these assets into fresh formats while maintaining brand consistency. Think of it as your digital brand manager, taking your established assets and adapting them for different audiences, channels, and objectives.

Producing excellent work means your prompting techniques are top notch. Next Up: Step-by-Step Guide to Great Prompts.

PART 8

DEEPER CHATGPT TRAINING

CHAPTER 18

Step-by-Step Guide to Great Prompts

You don't need to become a prompt engineer. You don't need to study the latest techniques. You just need to remember that you're giving instructions to something that's good at following directions, but terrible at reading your mind. Think of it like briefing a smart intern. The clearer you are, the better the result.

The Complete Formula

Here's a template you can use every time:

```
Act as [role/persona] and [do task] using [format].
The audience is [describe].
Keep it [constraints].
Use this [example] as a model.
Return the result as [format].
```

Example:

```
Act as a B2B conversion copywriter and rewrite this email
for first-time SaaS buyers.
The audience is skeptical operations leaders.
Keep it under 100 words and make the CTA stand out.
Use this previous email as a style guide.
Return the result as plain text.
```

Step 1: Be Clear About the Task

This sounds obvious, and it is. You'd be amazed how often people need to be reminded though.

"Make this better" isn't a task. "Rewrite this email to be more persuasive" is a task. "Turn this transcript into three key takeaways" is a task.

The AI isn't guessing what you want. It's pattern-matching based on your words. Give it something specific to match.

Good example:

Summarize this customer story into three key takeaways based on our value propositions.

Bad examples:

- "Make this better." (better how?)
- "What do you think of this?" (it doesn't think)
- "Can you work on this?" (work how?)

Step 2: Add the Right Context

Context is everything. Who's the audience? What's the goal? What tone are you going for?

I learned this the hard way. Early on, I'd ask for "marketing copy" and get generic fluff. Now I specify: "Write this for skeptical CFOs who care about ROI, not features."

Night and day difference.

Good example:

Rewrite this blog post for Ops Olivia. Keep it under 500 words and use a confident, insight-driven tone.

Why it works: Now the AI knows who it's writing for and how they prefer to consume information.

Step 3: Specify the Format

Structure matters. Bullets, paragraphs, numbered lists, tables. They all serve different purposes, and they change how the AI organizes its response.

If you don't specify, you're rolling the dice.

Good example:

Summarize this call transcript into 5 bullet points.

Each bullet should have a bolded title and 1 sentence of explanation.

Why it works: Clear structure makes the output immediately usable.

Step 4: Assign a Role

Give the AI a persona. This shapes everything: tone, priorities, knowledge base, approach.

"Analyze this landing page" could go anywhere. "Act like a conversion copywriter and analyze this landing page for clarity and persuasion" gives you focused feedback.

Good example:

Act like a B2B content strategist at a financial services firm.

Evaluate this landing page for clarity, persuasion, and conversion risk.

Step 5: Add Constraints and Goals

Constraints guide the AI away from fluff and toward usable output.

Word limits. Tone requirements. Specific objectives. These aren't restrictions. They're guardrails that keep everything on track.

Good example:

Write a subject line under 50 characters that highlights urgency and targets B2B SaaS buyers.

Step 6: Use Examples to Steer Quality

Show, don't just tell. If you have an example of what "good" looks like, include it.

AI is excellent at mimicking styles. Give it something to mimic.

Good example:

```
Here's a LinkedIn post I liked. Match its tone and structure
but write about our new feature instead.
```

Now let's look at the problems I see all the time. Next Up: Common Pitfalls to Avoid.

CHAPTER 19

Common Pitfalls to Avoid

Even with experience, it's easy to fall into traps. Here's what I see most often:

Being Too Vague

Vague prompts get vague results. Every time.

"Write a blog post" could mean anything. "Write a blog post introducing a new feature to mid-market SaaS marketers" gives the AI something to work with.

The more specific your intent, the sharper the result.

Overloading the Prompt

Trying to cram multiple tasks into one prompt usually backfires.

Instead of "Summarize, rewrite, analyze, and format this," break it into steps:

1. Summarize first
2. Then rewrite
3. Then analyze
4. Then format

Prompt chaining works better than prompt cramming.

Forgetting the Audience

Who is this for? If you don't specify, the AI fills in the blanks with guesswork.

"Write this announcement" could be for anyone. "Write this announcement for a skeptical CFO who cares about cost control" gives clear direction.

Not Recognizing When to Take Over

Sometimes the fastest way to finish is to stop prompting and start editing.

If you're spending more time tweaking prompts than it would take to rewrite the result, take the wheel. AI is a collaborator, not a replacement for judgment.

Trusting the Output Too Much

As you get comfortable with AI, it's tempting to believe everything it generates.

Don't.

The AI doesn't "know" anything. It predicts what sounds right based on training, not what is right.

Always verify quotes, stats, dates, and anything that sounds too good to be true. Remember, treat it like a clever intern. Check its work.

Next Up: Beyond ChatGPT - Tools Worth Exploring

PART 9

WHERE TO GO FROM HERE

CHAPTER 20

Beyond ChatGPT - Tools Worth Exploring

Let's zoom out. Here are tools worth knowing, broken into categories that actually matter for your work.

> ### Tip:
> *The AI landscape changes so fast, this list is out of date the moment you purchased this book. Use this as inspiration, not as a buying guide.*

Video: From Script to Clip

Video used to require teams, budgets, and weeks of back-and-forth. Now you can go from concept to clip in hours.

Descript turns transcripts into video edits. Upload a webinar recording, edit by cutting text, and you've got a social teaser in 15 minutes.

Pictory generates short-form videos from long-form content. Feed it a blog post, get back a video summary.

Runway handles advanced video editing with AI. Great for product demos that need visual polish.

Try this: Upload your last webinar to Descript. Cut three 60-second clips for LinkedIn. See how fast you can move.

Design: Visuals Without the Bottleneck

Need visuals but don't want to wait for design resources?

Midjourney creates images from text descriptions. The aesthetic leans artistic, which works well for concept visuals.

Canva AI generates images and suggests layouts for non-designers. Less artistic than Midjourney, more practical for business use.

Adobe Firefly integrates generative features into tools you might already use.

Try this: Use Canva's AI to generate five image variations for your next LinkedIn carousel. See which style fits your brand.

Audio: Voice Without the Studio

Create voiceovers or full podcasts without hiring voice talent.

ElevenLabs creates realistic AI voices from short samples. Great for consistent narration across content.

Wondercraft turns blog posts into podcast-style audio. Upload text, get back narrated content.

Try this: Convert your most popular blog post into a 10-minute audio version using Wondercraft. Test whether your audience prefers reading or listening.

Data + Research: Find Insights Faster

Move from search to synthesis without getting lost in rabbit holes.

Perplexity works like a research-focused search engine with source citations. Ask complex questions, get structured answers with links.

Claude excels at analyzing long documents and providing summaries. Great when you need to process dense material quickly.

Browse AI monitors websites for changes and updates. Set it up once, get alerts when competitors update pricing or features.

Try this: Use Perplexity to create a competitive landscape summary for your next board presentation. Include the sources so others can dig deeper.

Mental Models to Carry Forward

Here's what matters, regardless of which tools you choose:

AI is not magic. It's a collaborator. Don't outsource judgment. Use it to sharpen your thinking.

Your role is to shape the input. You don't need to be technical. You need to be clear.

Good inputs create leverage. Prompts, templates, and workflows aren't shortcuts. They're force multipliers.

Great marketers think like systems designers. The tools are your team. You decide how they work together.

Try one new tool this week, but bring the same lens you've used throughout this book: "How can this help me think, decide, or create faster?"

You don't need more tools. You need the right ones, used the right way. Be careful not to get so many AI subscriptions that you go broke paying for them.

Next Up: Your options for when you get stuck or want to level up faster.

CHAPTER 21

The Road So Far: Key Lessons and Next Moves

If you've made it this far, one thing's clear: you're serious about using AI to drive real business results. You don't have to travel alone; my company provides services designed to help you and your company with AI and product related topics. Check the first part of the appendix for more information.

Welcome Back to Day One

When you first opened this book, you felt like you were stepping into uncharted territory. The world of AI is overwhelming, especially for those who already juggle chainsaws and shards of broken glass. But here you are, armed with knowledge, tools, and frameworks to make AI work for you.

Remember the opening question we asked: *How can we use AI to simplify and elevate our marketing workflows?* The answer was about enhancing what you do, not replacing you. You've learned to use the power of AI to save time, be more creative, and eliminate inefficiencies in your process.

This chapter is here to help you take that final step toward making AI an integral part of your marketing journey.

Mindset Wins

It's easy to get caught up in the excitement and hype of new technology. The media portrays AI as a one-size-fits-all solution that will change everything overnight. We know better. AI is a tool, not a magic wand.

Through this book, you've embraced the mindset of a marketer who innovates with AI while keeping strategic focus intact.

You learned to balance automation with authenticity, understanding that AI's role is to simplify tasks, not erase your unique human touch. We use AI to augment our thinking with better data, more insights, and faster iterations.

Now, you are an AI-powered marketer who delivers high-impact content at scale, while keeping strategy at the forefront. Congratulations! This shift in mindset is your real superpower.

Foundations Built

Let's take a moment to look back at the foundational work we've done.

Company Brief, ICPs, Personas, and Brand Voice

We started by helping ChatGPT understand your business. You trained your assistant to produce content aligned with your goals and tone by setting up clear instructions, creating templates, and defining personas. This foundational pattern is the key to AI producing content that feels authentically *yours*, regardless of the task at hand.

Custom GPT Setup

You customized your AI assistant to work like a team member who's always in sync with your business objectives. This setup ensures that ChatGPT remembers your preferences, company values, and market nuances, producing content that meets your exact needs. It's a business tool that works *for* you, aligned with your unique work and workflow.

By focusing on these fundamentals, you've set yourself up for long-term success. The hard work of setting up the infrastructure now allows for easy expansion and fine-tuning.

Messaging that Holds Together

Good messaging is like good spaghetti sauce, thick and flavorful, not watery and bland. We started by reducing your core value proposition down to its flavorful essence, sprinkled in a set of messaging pillars for structure, and taste-tested the result across every channel in your mix.

Why the fuss? Because AI cooks only what you prep. Give it thin stock and you'll get thin copy. Give it a well-seasoned brief and the output sings. Think of your AI assistant as the world's fastest sous-chef: brilliant with a knife, but it still needs you to choose the ingredients and keep the salt under control.

From Strategy to Assets

You started this journey with a blank page and a hunch that AI could lighten the load. Now you have a repeatable system that carries a marketing idea from the first spark all the way to measurable results. Here's how we walked through each stage:

Campaign Strategy Ideation

- Use clear goals and well-defined personas to prompt AI for fresh angles.
- Treat its suggestions as draft fuel, keep the gems, toss the rest, and shape the big picture.

Asset Build-Out

- Turn that strategy into briefs, landing pages, email sequences, and ad copy in minutes, not weeks.
- Anchor every piece to your core messaging so the story stays consistent from first touch to final click.

Asset Recycling

- Feed existing blogs, videos, or slide decks back to the assistant.
- Let it remix and resize content for new channels, saving hours of manual rework.

Campaign Delivery and Execution

- Hand off polished assets to your marketing stack, like your CRM, ad platform, or webinar tool.
- Use AI to draft timelines, task lists, and even test plans so launch day feels calmer.

Post-Mortem and Insight Mining

- Pull performance data into ChatGPT for quick summaries of what worked and what missed the mark.
- Ask follow-up questions to surface patterns you might overlook, then log next-step ideas while they're fresh.

Continuous Refinement

- Cycle those insights back into the next ideation round.
- Let AI propose variants for you. Subject lines, offers, creative hooks should be iterated on. Then choose the strongest options for the next sprint.

To stretch the kitchen metaphor, you stay in charge of the recipe and final seasoning, while the bot handles the repetitive chopping. You'll get more dishes out the door, less kitchen chaos, and a marketing operation that keeps getting tastier with every pass through the loop.

Don't worry, you are still in control. You can do more than you could before and still maintain quality and awareness. Your closed-loop workflow is designed for strategy to feed assets, assets feeding execution, and execution feeding learning. You stay focused on creative decisions only a human can make.

Prompt Craft and Pitfalls

If you forget everything else in my book, remember this: great AI driven marketing starts with great prompts. Yes, it looks easy, type words into the magic content vending machine, watch clever copy pop out. But a lazy prompt is like cooking with no recipe: you might get dinner, you might get charcoal.

Over the last chapters, you built a toolkit of prompt tricks:

- **Prompt shaping** to steer tone and style.
- **Context building** so the AI knows the backstory.
- **Prompt chaining** to break big tasks into snack-size steps.

136

- **Diagnostic checks** to spot when the model is hallucinating a Nobel Prize it never won.

Clarity and specificity turn "meh" output into marketing gold. Keep your asks tight, your examples sharp, and your assumptions spelled out. When AI drifts, pull it back with a quick tweak instead of ten angry revisions.

Remember, you'll need to find the balance between when to tweak the prompt and when to edit by hand. It's more art than science.

Your New Operating System

AI is now part of your daily workflow, but power tools still need upkeep. Adopt these habits and you will stay ahead of the robots rather than behind their error logs.

Rhythm	Habit	Why It Matters
Daily	Skim everything your assistant produced. Tweak sloppy prompts on the spot.	Catching little issues now beats untangling a spaghetti mess later.
Weekly	Run a quick retro on live campaigns: what crushed it, what fizzled?	Fast feedback keeps your message fresh and your budget happy.
Quarterly	Step back. Audit new AI features, swap in better models, prune unused workflows.	Technology moves faster than your coffee cools. Stay current or get stuck.

Regular tune-ups keep the system humming and your sanity intact. Think of it like brushing your teeth: skip a day and nobody notices, skip a quarter and people start stepping back when you talk.

Next Steps

Pick one AI task today—maybe headline testing or persona research. Set a 72-hour deadline to improve it by one clear metric: speed, quality, or both. Run your new prompt, measure the lift, repeat. Small wins stack fast when a silicon sidekick does the heavy lifting.

Close the book, open your prompt window, and ship something better before lunch. The future favors the marketer who iterates.

Next Steps

Before we wrap up this journey, take one thing from this chapter and apply it immediately. Pick one AI-driven task you'd like to refine or expand. Set up a mini project, give it a deadline, and execute with the same level of intentionality you've used throughout this book.

Now, let's take a moment to reflect on everything you've learned and look ahead at the opportunities that await in the next chapter of this AI-driven world. Next Up: The Road Ahead: Thriving in an AI-Driven World

CHAPTER 22

The Road Ahead: Thriving in an AI-Driven World

As the AI landscape evolves, we participate in a revolution where smart, adaptable professionals like you will lead the way. The opportunities ahead are vast, and the pace of change will only increase for a while. But here's the good news: just as you've learned to harness AI to streamline your marketing workflows, you now have the tools and mindset to thrive in this rapidly changing environment.

In this chapter, we'll explore how you can continue to grow, adapt, and succeed in the world of AI-powered business, while keeping your human touch firmly in place.

A Quick Look Over the Horizon

If I could predict the future, I'd be writing this book from my private Caribbean Island. I'm not. I'm home in North Carolina. Given that, or perhaps in spite of that, I feel like I owe you a slice of my perspective. Take it for what it's worth.

In this new era, AI will be part of the fabric of business itself. If you look back at the evolution of technology, you'll see a consistent pattern: every technological advancement raised the bar, but it never made smart people irrelevant. In fact, each innovation increased the demand for skilled professionals who could harness these new tools for good. The skills changed, but there was never a fixed amount of need. Rather the market expectations increased to expect more quality, faster output, better content, more tailored messages, and so on.

AI follows that same arc. As AI continues to improve, the bar for performance will rise. So will the opportunities to use these technologies to make smarter decisions, execute faster, and provide better value to your customers.

Now, more than ever, the role of the human in the equation will be to innovate, adapt, and continue building.

7 Trends to Watch

The AI world is evolving at the speed of a toddler on espresso. Below are 7 areas I believe will be dramatically shaped by AI. For each, I sketch a market-level view and then drop you into a quick "day in the life" so you can picture the impact without a crystal ball.

Autonomous AI Agents Reshaping Workflows

Businesses are looking for tools that reduce manual oversight and improve both productivity and decision-making. Autonomous AI agents fulfill these needs by automating routine tasks and proactively offering insights.

Imagine you're the manager of a busy marketing team. Every morning, you spend the first hour of your day scheduling meetings, coordinating team tasks, and reviewing reports. It feels like a lot of time is lost on administrative tasks rather than focusing on creative strategies. Enter the autonomous AI agent, your new digital assistant.

This AI agent does more than respond to basic questions and build landing pages on command, it takes the right actions automatically. It schedules appointments based on team availability, monitors campaign performance in real-time, and even alerts you to potential risks before they escalate. Think of it as having a personal assistant who works 24/7, without needing constant direction. Not a replacement for the manager, rather an administrative machine that handles the repetitive tasks that keep the department moving forward. I could use one of these myself right now.

AI as an Invisible, Foundational Utility

The right amount of AI most people want to buy is zero. They actually want better outcomes. They crave shorter commutes, quicker check-ups, and Wi-Fi that behaves, without fiddling with technology. It just needs to work the way they want without breaking or otherwise requiring attention, so the tech that melts quietly into the background wins the popularity contest.

In smart cities, AI will ensure traffic lights are synced to reduce congestion. In healthcare, it'll help doctors make more accurate diagnoses, and in education, it'll adapt lessons to individual student needs. AI will become as ubiquitous as electricity, there when you need it, but never in the way. How often do you think about your electric company? Unless you work there, I bet you only think of them when there is an outage.

To be amazing it doesn't even have to be revolutionary, just in a current process that needs fast optimization on fuzzy variables. In a smart city, traffic lights adjust themselves every few seconds based on live camera feeds. A few more seconds during peak times would get more drivers through safely. Drivers would never think "What a brilliant tensor network," they'd just arrive on time and have road rage for a different reason, wouldn't they?

This isn't ready for us yet. Technology is unreliable at times and there are many questions of liability and fault that are yet to be clarified. However, it's coming.

Hardware Innovation & Energy Efficiency

Current chips are powerful but hungrier than a teenager at dinner. I've got 2 teenagers myself so this hits home. New AI-specific silicon will cut the electric bill and keep progress from hitting the physics wall.

See, currently, large AI models, particularly generative AI, demand immense computational resources, requiring significant investments in specialized hardware (like GPUs and custom AI accelerators). While state-of-the-art hardware is powerful, the scale of AI pushes beyond readily available and affordable computing power. Plus, all this hardware needs electricity and all of a sudden we must bring many new power generation stations online to satisfy the AI needs.

Moore's Law has stated for decades that the performance of microchips will double every 18-24 months, but this has been slowing down recently. The rate of improvement in transistor density and general-purpose CPU performance has slowed due to fundamental physical limits and the increasing complexity and cost of manufacturing at atomic scales. We are already manufacturing chips using extreme ultraviolet (EUV) lithography, a technology that uses light generated artificially in a vacuum (Google it, it's cool!). There's little room left for further scaling with current materials and architecture.

I bet using special-purpose AI, we'll discover newer ways to manufacture chips that can handle AI workloads more naturally and with less power consumption. While performance for AI workloads on specialized hardware is currently improving rapidly, whether this constitutes a sustained "new Moore's Law" for AI performance remains speculative.

I hope to see it myself; but the ongoing research into AI-specific architectures and advanced manufacturing techniques suggests big future gains at least. Whether we get a "Moore's Law 2.0" or just "Moore's Suggestion" remains to be seen, but nobody's betting on coal-fired GPUs, are they?

Maturing AI Governance & Responsible AI Practices

As AI becomes more integrated into sensitive applications, companies must establish clear governance and ethical guidelines to build trust and ensure regulatory compliance. This includes conducting regular audits, maintaining transparency in data use, and committing to ethical AI practices.

The Wild West phase is ending. Regulators are rolling into town, clipboards in hand, asking who owns the data, who authorized certain tools, and whether the chatbot read the privacy policy.

Data privacy is not limited to personal data. It extends also to business data. As a result, companies will increasingly control which AI products can be used and in what contexts, to safeguard sensitive information from misuse.

The modern CISO is both excited and cautious about the evolving risks in this space. There are security implications that have yet to fully materialize. We can expect the compliance slide in board decks to grow from one bullet to several lovingly detailed pages, crippling the ambitions of early adopters everywhere.

AI-Driven Scientific Discovery in Healthcare

Health diagnostics used to sound like, "Come back next week and we'll know more." Today it's edging toward, "Grab a coffee, your genome will be ready in five minutes." AI now scans medical images, genetic data, and lab results in a fraction of the time it once took, turning weeks of anxious waiting into same-day answers. Faster, more precise diagnoses free clinicians to spend their energy on treatment and bedside care instead of spreadsheet archaeology.

The same pattern-spotting muscle powers public-health surveillance. Machine-learning models sift electronic health records, social-media chatter, environmental sensors, and genomic feeds to flag outbreaks before they hit the evening news. During the flu season, or heaven forbid, the next COVID-style surge, AI will forecast hot spots, help officials move resources sooner, and ultimately save more lives.

What's next? Personalized medicine at scale. Right now, humans are treated as a group, but disease and circumstance are individualized. If we can analyze individual health data efficiently, we unlock personalized medicine for the masses.

Once AI can crunch an individual's full data trail, genes, lifestyle, even the air quality on their morning jog, care shifts from "average best guess" to "tailored just for you." Expect the next big leap in health to be personal, precise, and powered by algorithms that never need a coffee break.

I expect AI to play a big role in personalized care. I'm really looking forward to this one.

AI-Driven Molecular Discovery in Pharmaceuticals and Materials Science

AI is increasingly being used to speed up molecular discovery, particularly in pharmaceuticals and materials science. The modeling software chemists once nursed for weeks now screens millions of molecules between dinner and breakfast. In pharma, these models predict how compounds behave in the body, flagging the best drug candidates in days rather than years. I imagine a researcher kicking off an antiviral screen at 6 p.m. and arriving the next morning with a shortlist of hits and a valid excuse to scrap tomorrow's marathon lab meeting.

The same horsepower works in materials science. Algorithms digest decades of test data, then suggest battery chemistry, metal alloys, or catalysts that human intuition would probably overlook. Teams skip the "mix and pray" stage and go straight to high-odds winners.

As demand increases and materials become scarcer and more expensive, we need new ways to build things.

The payoff is faster, cheaper innovation everywhere from energy storage to aerospace. When AI handles molecular matchmaking, scientists focus on the fun part: turning discoveries into real-world breakthroughs.

Geopolitical Competition and Supply Chains

Geopolitics is the weather forecast no one can ignore. It's unpredictable, far-reaching, and prone to sudden storms. To stay competitive, companies need a live radar on shifting trade rules, export bans, and chip shortages, plus a plan B for every "just in case."

Given the complexity of supply chains and geopolitical factors, even an army of Albert Einsteins would struggle to process and make decisions effectively. This is where AI earns its keep. Risk-modeling systems can crunch tariffs, shipping data, and political chatter in real time, then spit out "If X happens in Country Y, here are three supply-chain detours that won't tank the quarter." Let the algorithms handle the global chessboard; you focus on the next strategic move.

Adopt the Perpetual Learner Mindset

If there's one thing you need in this evolving AI world, it's the mindset of a perpetual learner. Technology moves fast, and the best way to keep up is by embracing curiosity. Here's how you can do it:

Skill Stacking

Pair your domain expertise with AI fluency. For example, you could deepen your understanding of customer psychology while learning how to prompt AI for more meaningful insights.

Experiment Often

The best way to learn is by doing. Don't wait for the perfect solution—start experimenting with AI tools today. You'll learn faster by iterating and improving as you go.

First Principles Curiosity

Think like a beginner. Ask the foundational "why" questions. Why does this AI tool work this way? Why do certain types of prompts get better results? This mindset will help you understand the core mechanics of AI, rather than just using it as a black box.

Opportunities for Smart People Who Show Up

While AI is transforming the landscape, it's also creating exciting new opportunities for those who are ready to take advantage of them. Smart people—like you—will be in demand because you know how to work with AI, not just rely on it. Here are some new roles and opportunities that are emerging:

AI Workflow Designers: Professionals who specialize in creating efficient AI-assisted workflows. This role will be crucial as businesses start integrating AI more deeply into their operations.

Model Ethicists: As AI systems make more decisions, the need for professionals who understand AI ethics will grow. These individuals will help ensure that AI is used responsibly, ethically, and transparently.

Prompt Librarians: A unique role where individuals create and manage libraries of effective prompts for different use cases—marketing, sales, customer service, and more.

145

AI will create new jobs and also help existing professionals level up. Your future job might not be "AI expert", but you will be someone who knows how to integrate AI tools into your work seamlessly.

Responsible AI in Practice

With power comes responsibility. As we become more reliant on AI, it's important to understand the ethical implications and ensure that we use AI in a way that aligns with our values. Here are some key considerations:

Data Privacy

Ensure that the data you use to train AI is ethically sourced and used in compliance with privacy laws.

Bias in Models

AI models can reflect biases in their data. Regularly audit your models for fairness, ensuring they don't inadvertently harm certain customer segments.

Transparency

Be clear with your customers and stakeholders about how AI is being used. Transparency builds trust and helps mitigate the risks associated with black-box decision-making.

By keeping these ethical considerations front-and-center, you'll not only safeguard your business from potential pitfalls but also be seen as a leader in responsible AI use.

Your 90-Day Growth Plan

As we conclude this chapter, it's time to look at your action plan. Here's a simple 90-day framework for incorporating AI into your workflows and growing with it:

Week 1-4

Audit one workflow in your business and experiment with a new AI tool. Make sure it fits seamlessly into your process.

146

Week 5-8

Gather metrics on the AI-generated content and the efficiency of your processes. Start refining your prompts and outputs for greater precision and quality.

Week 9-12

Teach a peer how to incorporate AI into their workflow. Share your insights and expand your efforts to other areas of the business. Begin exploring new tools that might offer more opportunities for growth.

Final Call to Action

AI is here to stay, and so are you. The tools will change, new models will emerge, and your knowledge will continue to grow. One thing will remain constant: the need for smart, creative, and adaptable professionals who know how to work with AI.

So, take a deep breath and think about the possibilities ahead. Stay curious. Stay brave. And remember, the future is yours to shape with the power of AI. As we've seen, the opportunity for growth is endless. You're already part of that future. Keep building, experimenting, and learning, and you'll continue to thrive in this AI-driven world.

APPENDIX

Part 1 – How to get Help

At DataCurl, we're not another firm handing you a 50-page playbook you'll never use. We're operators who've built things, scaled teams, and made the mistakes so you don't have to.

We partner with ambitious companies to turn AI from buzzword into bottom-line results.

What We Actually Do

We specialize in AI consulting and go-to-market strategy for companies that don't have time to waste.

Custom AI Workflows & GPT Builds Build internal assistants that actually get used. Streamline customer support without losing the human touch. Automate the repeat work that's draining your team's energy.

AI-Powered Product Marketing & Management Whether you're launching an AI product or adding AI features to an existing platform, we help with strategy, messaging, and user adoption. The stuff that determines whether your AI features get used or ignored.

Strategic Consulting for AI Features Thinking of adding AI to your product? We help product and engineering teams scope, design, and roadmap features that deliver ROI instead of just checking boxes.

Go-to-Market Systems From positioning and pricing to email sequences and outbound plays, we create repeatable, AI-informed systems that accelerate growth without adding headcount.

Team Training & Workshops We'll teach your team to think in prompts, automate workflows, and build internal AI expertise. Fast, practical, immediately useful.

Who We Work With

We partner with venture-backed startups that need to move faster than their competitors. PE-owned firms that need results, not reports. Traditional

150

businesses modernizing with AI. Niche product companies refining their go-to-market approach.

If you're looking for high-impact work delivered by people who've actually built things, we should talk.

Ways to Work with Us

Book a 1:1 Strategy Call: Fast, focused, actionable. Get expert insight in a single conversation.

Build Your Custom AI Assistant: Co-create a GPT-powered internal tool with prompt libraries tailored to your specific workflows.

Product or GTM Engagement: From AI roadmap strategy to full-stack go-to-market systems. We do the work with you, not for you.

Need help building your AI assistant or launching your next product?

Visit datacurl.com to book a free discovery call.

You can also connect with me on LinkedIn. It's good to know who you're building with.

Let's move fast and build smart.

Part 2 - Marketing Worksheets

> *Tip:*
>
> *Get downloadable worksheets at datacurl.com/books.*

Minimum Viable Company Brief Worksheet

Copy, print, or paste into your notes. Each bullet is a field to complete.

1. Snapshot – "Who are we?"

One-line identity (industry | category | location):

_____

Founding year & stage (idea, pre-seed, seed, Series A, etc.):

_____

Mission in a sentence:

_____

2. Core Offering – "What do we sell?"

Flagship product or service name:

_____

Primary value (outcome delivered):

_____

Pricing model (one-time, subscription, usage-based, other):

_____

3. Target Customer – "Who do we serve?"

Ideal segment (ICP name or market slice):

_____

Key persona / buyer title:

_____

Pain they feel today (one sentence):

_____

4. Need Timing – "When do they look for us?"

Typical triggering event (e.g., funding, audit, churn spike):

_____

Urgency window (immediate, 90 days, annual cycle):

_____

5. Problem–Solution Fit – "How do we help?"

Top three pains we solve:

1) _____

2) _____

3) _____

Top three benefits we deliver:

1) _____

2) _____

3) _____

6. Strategic Why – "Why does it matter?"

Bigger impact (vision or purpose):

_____

Proof point or stat that supports the impact:

_____

7. Messaging Pillars – "What must every pitch reinforce?"

One-sentence explanation

1) _____

2) _____

3) _____

8. Signature Use Cases – "How customers apply us"

Outcome in one phrase

1) _____

2) _____

3) _____

9. Competitive Landscape – "Who else is in the room?"

Direct alternatives (same category):

_____

_____

_____

Workarounds currently used:

_____

_____

_____

Differentiator in five words:

_____

_____

_____

10. One-Paragraph Company Blurb

Write 3–5 sentences that stitch the answers above into a story.

_____

_____

_____

_____

_____

_____

Guidance for Each Section + ChatGPT Prompts

Worksheet Section	What to Think About	ChatGPT Starter Prompt
Snapshot	New hires and collaborators need a crisp headline. Keep jargon out and clarity in.	"Draft a one-line identity for a seed-stage SaaS that automates B2B onboarding."
Core Offering	Focus on ONE thing you ship today, not the five in your backlog. Tie value to an outcome.	"Rewrite our product description so a non-technical founder sees the core benefit in 25 words."

Target Customer	Lean on your earliest paying users. Name the persona, not the company logo.	"Summarize the common traits of the first ten customers who paid for our beta."
Need Timing	Map the buying moment. If you know the trigger, you can time campaigns and outreach.	"List three business events that push a Head of Ops to seek a workflow automation tool."
Problem–Solution Fit	Rank pains by cost or frustration level. Limit each to one line to stay sharp.	"Turn the support tickets below into a ranked list of problems we solve."
Strategic Why	This connects your product to a mission bigger than profit. A compelling 'why' rallies both staff and buyers.	"Convert our product vision into a single sentence that passes the 'so what' test."
Messaging Pillars	Pillars become themes for website copy, decks, and ads. Three is memorable; more is noise.	"Propose three messaging pillars for an early-stage AI document-review platform, each <10 words."
Signature Use Cases	Pick use cases that buyers already search for. Show variety: one strategic, one operational, one quick win.	"Suggest three high-impact use cases for manufacturers adopting our predictive maintenance API."
Competitive Landscape	Honest awareness builds credibility. List real rivals	"Create a competitive snapshot contrasting us with the top two no-code

	plus the 'do nothing' status quo.	AI agents and with spreadsheets as the workaround."
One-Paragraph Blurb	This is your default PR paragraph and ChatGPT context block. Make it scannable and punchy.	"Using the filled worksheet, craft a 120-word company blurb suitable for our website footer."

Tip:

After filling the worksheet, paste it into ChatGPT and ask:
"Evaluate this brief for unclear jargon or missing proof points.
Suggest improvements line by line."

Iterate until every section is specific, believable, and inspiring. Then save the brief as the go-to reference for all future marketing tasks.

Ideal Customer Profile (ICP) Worksheet

Tip:

Get downloadable worksheets at datacurl.com/books.

Print or copy into your notes. Each bullet marks a field the student should complete.

1. Firmographics – "Who are they?"

Primary industry / niche:

_____

Annual revenue range:

_____

Employee headcount:

_____

Geography / service regions:

_____

Ownership & growth stage (bootstrapped, VC-backed, PE-owned, public, etc.):

_____

2. Technographics – "How do they work today?"

Core software stack (ERP, CRM, cloud, etc.):

_____

_____

_____

AI / automation maturity (early, experimenting, scaling, advanced):

_____

Must-have integrations or compliance requirements:

_____

_____

_____

3. Business Model & Structure – "How do they make money & buy?"

Model (B2B, B2C, marketplace, other):

_____

_____

Go-to-market motion (sales-led, product-led, channel, hybrid):

_____

_____

Buying process (centralized, committee, self-serve):

_____

_____

4. Pain Points & Challenges – "What's broken?"

Top operational or strategic pain (1):

_____

_____

Secondary pain (2):

_____

_____

Current workaround / cost of inaction:

_____

_____

5. Goals & Objectives – "What are they trying to achieve?"

Primary business goal your product supports:

_____

_____

_____

Success timeline (e.g., 6-month target):

_____

_____

_____

Strategic initiatives already in play:

_____

_____

_____

6. Buying Triggers – "Why now?"

Events that spark urgency (funding, audit, leadership change, regulation):

_____

_____

_____

Symptoms that appear internally (missed KPI, rising churn, etc.):

_____

_____

_____

7. Success Criteria & KPIs – "How will they measure value?"

Quantitative metric (time saved, revenue gained, cost reduced):

_____

_____

Qualitative win (brand, morale, risk reduction):

_____

_____

_____

ROI hurdle / payback period expected:

_____

_____

_____

8. Ideal Buyer Persona(s) – "Who signs the deal?"

Role	Typical titles	Motivation	Likely objections
Champion			
Decision-maker			
Influencer / Gatekeeper			

9. Readiness To Purchase Score – quick gut check

Fit (1–5)	Need (1–5)	Budget (1–5)	Timing (1–5)	Total /20

10. One-Paragraph ICP Snapshot

Write 3-5 sentences that capture the essence of this ideal customer.

_____

_____

_____

_____

_____

How to Think Through Each Section (with ChatGPT jump-start prompts)

Worksheet Section	What to Consider	Starter Prompt for ChatGPT
Firmographics	Where does your earliest traction appear? Focus on the size and sectors that buy quickly, not the dream logo list.	*"Given our pilot customers in SaaS and logistics, what common firmographic traits do you see (industry, size, geography)?"*
Technographics	List systems your product must connect with – if you integrate smoothly, you shorten sales cycles. Note their appetite for new tech (early vs. late adopters).	*"Create a table showing the typical tech stack for a mid-market logistics firm adopting AI, and flag gaps our product can fill."*
Business Model & Structure	Map how they sell and how they buy. A product-	*"Summarize how a product-led SaaS*

	led SMB behaves very differently from an enterprise with a buying committee.	*company under 200 employees usually evaluates a workflow tool like ours."*
Pain Points & Challenges	Interview notes, support tickets, and competitor reviews are gold. Prioritize pains that are urgent *and* expensive.	*"List the top three operational pains for a VC-backed SaaS COO trying to scale without new headcount."*
Goals & Objectives	Tie your benefit to a goal they already track (ARR, churn, deployment speed). If you can't anchor to a KPI, you'll struggle to prove value.	*"Rewrite our product benefit in terms of metrics a COO reports to the board."*
Buying Triggers	Triggers turn a "someday" problem into a "this-quarter" budget. Think funding rounds, new compliance rules, tech sunsets.	*"What external events typically push a mid-market manufacturer to seek an AI workflow solution?"*
Success Criteria & KPIs	Ask prospects how they'll judge success before you sell. Capture both hard numbers and softer wins like team morale.	*"Draft three measurable success criteria a RevOps leader would cite after deploying our platform."*
Ideal Buyer Persona(s)	Separate the *champion* (day-to-day user), the *economic buyer*, and any blockers. Note each one's personal win and fear.	*"Based on this ICP, outline champion, decision-maker, and blocker personas with*

		their motivations and objections."
Readiness To Purchase **Score**	Scoring helps reps disqualify politely and focus on high-probability deals. Define what a 5 vs. a 3 looks like for each axis.	*"Suggest a simple 1-5 rubric for assessing Fit, Need, Budget, Timing for early-stage SaaS prospects."*
ICP Snapshot	This is the elevator pitch for your internal team and for ChatGPT later. Make it vivid: "A 200-person SaaS firm, Series B, struggling to…"	*"Condense the completed worksheet into a 100-word narrative any new hire can grasp."*

Pro tip:

Once you draft an ICP, paste the filled worksheet into ChatGPT and ask: "Challenge this ICP: what assumptions look risky for an early-stage company, and what data should we gather next?"

Use the AI as a sounding board but ground every answer in real conversations with prospects. The tighter your inputs, the sharper your marketing.

Key Buyer Persona Worksheet

Tip:

Get downloadable worksheets at datacurl.com/books.

---

*Copy, print, or paste into your notes. Each bullet is a field to complete for **one** persona. Create 2-4 separate worksheets if you have multiple buyer roles.*

1. Persona Snapshot – "Who are they?"

Alliterative name (e.g., Ops Olivia):

_____

Real-world title(s) & level (VP, Director, Manager):

_____

Department / function:

_____

Adoption role (Champion, Decision Maker, Influencer, Skeptic, End User):

_____

2. Goals & KPIs – "What keeps them measured?"

How success is tracked

1) _____

2) _____

3) _____

3. Job Responsibilities – "What do they do each week?"

Top recurring tasks (bullet list):

_____

_____

_____

_____

_____

4. Pain Points & Active Problems – "Where does it hurt?"

Business impact

1) _____

2) _____

3) _____

5. Buying Motivations – "Why would they say yes?"

Personal wins:

1) _____

2) _____

3) _____

Team or company wins:

1) _____

2) _____

3) _____

6. Objections & Concerns – "Why might they hesitate?"

Primary objections:

1) _____

2) _____

3) _____

Secondary objections:

1) _____

2) _____

3) _____

7. Research & Evaluation Habits – "How do they shop?"

Channel (peers, analyst, webinar, etc.)	Trust level (high / medium / low)

8. Influencers & Stakeholders – "Who shapes the decision?"

Role	Influence on deal	What they care about

9. Communication Style – "What resonates?"

Preferred tone (data-driven, visionary, tactical):

_____

Content formats they prefer (case study, demo, checklist, etc.):

_____

_____

10. Internal Wins – "How do they look good?"

Quick win they can show	Metric or story shared upward

11. Success Metrics Post-Purchase – "What proves you delivered?"

Metric	Target value	Timeframe

Notes:

_____

_____

_____

_____

How to Fill Each Section (with Starter Prompts)

Worksheet Section	What to Think About	ChatGPT Starter Prompt
Persona Snapshot	Choose a memorable nickname and gather the most common job titles you meet in sales calls.	"List common titles and a memorable alliterative nickname for the internal AI champion within a mid-market SaaS firm."
Goals & KPIs	Tie goals directly to metrics already on their bonus plan. Revenue, churn, and cycle time beat vague "innovation" talk.	"Based on recent LinkedIn posts from VPs of RevOps, what KPIs do they cite most often as success measures?"

Job Responsibilities	Scan job ads and LinkedIn profiles to capture everyday tasks. This grounds your messaging in reality.	"Summarize the weekly responsibilities of a Director of Customer Success at a Series B SaaS company."
Pain Points & Active Problems	Rank pains by business impact and urgency. Ignore annoyances that will not justify budget.	"Convert these support-call notes into a ranked list of pains for a logistics COO, including the cost of each pain."
Buying Motivations	Separate personal career wins from company outcomes. Both matter in B2B deals.	"Describe the personal and company-level motivations that push a CIO to buy workflow automation."
Objections & Concerns	Pull from real sales calls. Address hidden fears like job security or change fatigue.	"List three likely objections a CFO raises when assessing a no-code AI platform priced on usage."
Research & Evaluation Habits	Identify where they look first – peer Slack groups, Gartner, Reddit, podcasts. This informs channel strategy.	"Map the typical research journey for a Head of Ops comparing AI vendors in manufacturing."
Influencers & Stakeholders	Chart who can veto or boost the purchase – IT, finance, legal. Note their hot buttons.	"For this persona, list the three internal stakeholders most likely to influence their decision and what each one prioritizes."

Communication Style	Match tone to persona. A visionary likes big-picture stories; a skeptic wants proof and numbers.	"Suggest the tone, evidence types, and content formats that resonate with a detail-oriented IT security lead."
Internal Wins	Frame quick wins they can show their boss within 30–60 days.	"Give two fast wins a RevOps leader can achieve in the first month after deploying our tool."
Success Metrics Post-Purchase	Define measurable outcomes you can commit to in case studies later.	"Draft three measurable success metrics a digital transformation leader would track after rolling out our platform."

Next step: After completing one worksheet, paste it into ChatGPT and ask: "Challenge this persona profile for assumptions or missing data. Suggest follow-up questions to validate each section."

Iterate until each persona feels like a real person on your next sales call. Save the set alongside your ICPs and Company Brief so every future marketing asset can name-check "Ops Olivia" or "Tech Exec Tom" and speak their language.

Brand Voice & Style Guide Worksheet

Tip:

Get downloadable worksheets at datacurl.com/books.

Copy, print, or paste into your notes. Each bullet is a field to complete.

1. Brand Snapshot – "Who is speaking?"

Mission in one sentence:

Brand archetype (trusted guide, bold innovator, friendly coach, etc.):

Primary audience segment(s):

2. Tone of Voice – "How should we sound?"

Attribute	Primary ✓	Secondary ✓
Confident		
Friendly		
Authoritative		
Conversational		
Technical		
Warm		
Humorous		

3. Personality Filters – "What feels on-brand?"

Three descriptive adjectives (e.g., pragmatic, optimistic, candid):

1) _____

2) _____

3) _____

One sentence that captures the vibe:

4. Reading Level & Clarity

Target grade level (e.g., 8-10):

Complexity notes (define jargon or avoid it?):

_____

_____

_____

5. Writing Style Rules – "Mechanics that never change"

Rule	Example
Paragraph length	1-3 short sentences
Voice	Active ("We build workflows")
Preferred perspective	First person plural ("we/our")
Sentence variety	Mix short and medium
Bullet usage	For 3+ related ideas

6. Language Do's and Don'ts

Do	Don't
Use plain English	Use jargon such as "synergize"
Show evidence with numbers	Over-promise with hype
Ask clarifying questions	Assume facts not supplied
	Add emojis or slang
	Start with clichés ("In today's world…")

7. Vocabulary & Phrase Bank

Preferred terms (list 5-10):

1) _____

2) _____

3) _____

4) _____

5) _____

Banned words/phrases:

_____

_____

_____

8. Formatting Preferences

Headings style (Title Case? Sentence case?):

_____

CTA format (imperative, ≤6 words):

_____

Punctuation quirks (avoid em dashes, use Oxford comma, etc.):

_____

9. Anti-AI Tells to Eliminate

Common tell	How we avoid it
Overlong intros	Lead with the value
Repetition of key phrases	Vary wording
Passive filler	Replace with action verbs

10. Content Priorities – "Order of importance"

1. Reader outcome or benefit

2. Current pain point

3. How we solve it

4. Proof or example

5. Next step (CTA)

11. Review Checklist – "Before we publish"

o Tone matches attributes in Section 2

o Reading level goal met

o No banned words or AI tells

o CTA present and clear

How to Complete Each Section (with ChatGPT Prompts)

Worksheet Section	What to Consider	ChatGPT Starter Prompt
Brand Snapshot	Summarize brand intent and audience in one breath.	"Using our Company Brief, draft a one-sentence mission and name a fitting brand archetype."
Tone of Voice	Choose 1-2 traits that dominate, plus 2-3 supporting notes.	"List three tone attributes that resonate with Ops-level buyers in mid-market SaaS."
Personality Filters	Adjectives should cue style decisions.	"Suggest three adjectives that capture a confident yet approachable AI vendor."

Reading Level & Clarity	Lower grade levels increase reach without dumbing content down.	"Rewrite this paragraph at a 9th-grade level while keeping technical accuracy."
Writing Style Rules	Lock rules you wish every freelancer knew.	"Turn our preferred writing habits into a rules table with examples."
Language Do's and Don'ts	Banned words save editing time later.	"Create a 'do versus don't' list comparing plain English to buzzwords for our brand."
Vocabulary & Phrase Bank	Keep terms that reinforce positioning.	"From recent case studies, pull out recurring phrases that feel on-brand and list five we should keep."
Formatting Preferences	This governs headings, bullets, punctuation quirks.	"Propose formatting guidelines for blog posts aimed at technical executives."
Anti-AI Tells	Note patterns you delete when editing AI drafts.	"Identify common AI writing quirks in this sample and suggest fixes."
Content Priorities	Teach writers the hierarchy of info.	"Outline the five-step flow that hooks a CIO in the first 100 words."
Review Checklist	Minimal list you can run through in 60 seconds.	"Create a publication checklist based on our voice guide sections."

Next step: After filling the worksheet, paste it into ChatGPT and ask: "Audit this Brand Voice Guide for contradictions or gaps. Propose any clarifications needed."

Once refined, store the guide alongside your Company Brief and Personas. Every future prompt to ChatGPT should reference these three assets to keep content consistent and unmistakably yours.

Custom GPT Configuration Worksheet

> ### *Tip:*
> *Get downloadable worksheets at <u>datacurl.com/books</u>.*

Copy, print, or paste into your notes. Fill one worksheet for each assistant you create.

1. GPT Identity

Name (plain, descriptive)

_____

Short description (20 words or fewer)

_____

_____

_____

_____

2. Role & Purpose

One-sentence job statement

_____

_____

3. Core Knowledge Pack

List the files or text you will upload or paste (✓ when included)

- Company Brief ✓
- Buyer Personas ✓
- ICPs ✓
- Brand Voice Guide ✓
- Other

_____

_____

_____

_____

_____

4. Tone & Style Summary

Key traits from your Brand Voice (maximum three)

1) _____

2) _____

3) _____

5. Behavior & Workflow Rules

Tick the boxes that apply and add extras as needed.

□ Ask clarifying questions

□ Restate user request first

□ Require persona selection if missing

□ End with next-step suggestion

Additional rules:

_____

_____

_____

6. Things to Avoid

(List banned items such as emojis, legal advice, etc.)

_____

7. Capability Toggles

Check capabilities to enable.

☐ Web browsing

☐ Code interpreter

☐ DALL·E

☐ File uploads

_____

_____

_____

8. Starter Prompts (optional)

Prompt 1: _____

Prompt 2: _____

Prompt 3: _____

9. Sample Interaction

Use this to test the assistant after setup.

Q: _____

A: _____

_____

_____

10. Publish Meta

Icon / image source: _____

Visibility (private or share link): _____

Last updated: _____

Fill out each section, then transfer the details into the Custom GPT builder to complete your configuration.

How to Complete Each Section (with ChatGPT Prompts)

Worksheet Section	What to Consider	ChatGPT Starter Prompt
GPT Identity	A clear name beats clever branding in the sidebar. The description shows in the picker—keep it utility-focused.	"Suggest three descriptive names and one-line summaries for a GPT that writes product-led case studies."
Role & Purpose	Distill why this GPT exists. One sentence guides every response.	"Condense this detailed job description into a single 'role & purpose' line for the assistant."

Core Knowledge Pack	Load only what the GPT must reference often. Fewer but sharper docs improve accuracy.	"Given these five brand docs, which should be primary knowledge files for a marketing assistant GPT and why?"
Tone & Style Summary	Pull top traits from your Brand Voice Guide so the GPT stays on-brand even in short replies.	"Extract the three most critical tone cues from our voice guide."
Behavior & Workflow Rules	Think through common misfires: incorrect persona, missing CTAs, invented data. Bake safeguards here.	"List five practical behavior rules that stop a marketing GPT from fabricating metrics."
Things to Avoid	Add any recurring edits you make (excessive hype, em dashes, emojis).	"Create a 'never do' checklist for an enterprise-facing tech brand."
Capability Toggles	Only enable what you trust. If accuracy is critical, you might leave browsing off.	"Recommend which capabilities to enable for a GPT writing gated-content outlines, with pros and cons."
Starter Prompts	Quick-start buttons help teammates discover value fast. Pick tasks you perform weekly.	"Suggest two starter prompts that show off our GPT's strengths to new users."
Sample Interaction	Write a test question that touches core files and style rules. Use the	"Draft a short Q&A exchange that would confirm the GPT applies

	answer to verify tone and accuracy.	our persona and style correctly."
Publish Meta	A distinctive icon speeds recognition when your GPT list grows. Decide if the link should stay private or be shareable.	"List three minimal icon ideas for a GPT focused on logistics ops content."

Next step: After completing the worksheet, paste each filled section into the Custom GPT builder fields, upload the knowledge files, run your sample interaction, and iterate until the reply feels like it came from one of your team members.

Messaging Pillars Development Worksheet

> *Tip:*
>
> *Get downloadable worksheets at datacurl.com/books.*

Print, duplicate, or copy into your notes. Each blank line or table cell is a space to write.

1. Context Snapshot

Primary objective for these pillars (e.g., website relaunch, new-feature launch)

_____

_____

Chosen messaging framework (circle)
JTBD / StoryBrand / Challenger / Golden Circle / PAS / AIDA / 4 Ps / FAB / SPIN / Before-After-Bridge / Other _____

Key personas this set must speak to (list 2–3 nicknames)

Persona 1: _____

Persona 2: _____

Persona 3: _____

2. Core Value Proposition (one crisp sentence)

What overarching promise unites every pillar?

_____

_____

3. Draft Messaging Pillars

Aim for 3–5 pillars. Each one should state a benefit, not a feature.

Pillar #	Pillar statement (≤ 12 words)	Primary persona served	Business proof / evidence
1			
2			
3			
4			
5			

4. Proof Point Brain-Dump

List stats, case snippets, or anecdotes you can attach beneath each pillar.

Pillar 1 Proof Ideas:

_____

_____

Pillar 2 Proof Ideas:

_____

_____

Pillar 3 Proof Ideas:

_____

_____

Pillar 4 Proof Ideas:

_____

_____

Pillar 5 Proof Ideas:

_____

_____

5. Resonance & Coverage Check

Check each box that applies to the full pillar set.

Test	✓
Every core persona sees one pillar written for them	
At least one pillar addresses an emotional win	
At least one pillar addresses a rational / ROI win	
Each pillar can be backed by specific proof	
Language matches Brand Voice guidelines (tone, reading level)	

Gaps noted:

_____

_____

_____

_____

_____

_____

6. Ladder-Down Copy Variations

Write shorter forms that ladder down from the pillars.

	Copy (fill in after drafting)
10-word tagline	
25-word elevator pitch	
50-word blurb	
100-word paragraph	
250-word passage	

7. Peer / AI Feedback Loop

Questions to ask ChatGPT or a colleague once draft is filled:

1. "Rewrite Pillar __ to be sharper and more buyer-centric."

2. "Suggest one proof point for each pillar drawn from our case studies."

3. "Check this pillar set against Persona __'s top objection—what's missing?"

Notes from feedback:

_____

_____

_____

8. Final Sign-Off

• **Approved pillar set version #:** _____

• **Date:** _____

• **Next deployment (asset & owner):**

_____

_____

How to Use This Worksheet

1. **Fill Sections 1–2** to lock purpose and POV.

2. **Draft 3–5 pillars** in Section 3—keep them crisp and benefit-led.

3. **Attach proof** in Section 4 so every claim is defensible.

4. **Run the resonance check**; rewrite until every box is ticked.

5. **Create short forms** in Section 6 for taglines, bios, and ads.

6. **Get feedback** (Section 7) from ChatGPT and humans, iterate, then record the final set in Section 8.

Complete this worksheet and you'll have messaging pillars sturdy enough to anchor campaigns, sales decks, and AI-generated content, without drifting off-brand or off-target.

AI-Powered Campaign Planning Worksheet

Tip:

Get downloadable worksheets at datacurl.com/books.

Fill one worksheet per campaign. Duplicate the page for future launches.

1. Campaign Snapshot

Working title / theme

_____

Primary goal (one sentence & numeric target)
e.g., "Book 50 demo calls in 6 weeks."

_____

_____

_____

Core persona focus (nickname + role)

_____

_____

Key message or tagline anchoring the campaign

_____

_____

2. Strategy Alignment Check

Tick each box you can answer confidently. Circle gaps you must clarify before moving on.

Question	✓
Goal is measurable and time-boxed	
Persona's pain & desired win are clear	
Message aligns with at least one messaging pillar	
Budget / resources confirmed	
Success metric agreed by sales & marketing	

3. Channel & Tactic Brainstorm

List up to **5** primary channels where your persona actually spends time. Add a key tactic for each.

Channel	Primary tactic	KPI
1		
2		
3		
4		
5		

4. 6-Week Timeline Planner

(Adjust length as needed. Shade the cell where an activity lands.)

Week →	0 (Prep)	1	2	3	4	5	6
Asset creation							
Teaser / awareness							
Lead magnet launch							
Live event / webinar							
Follow-up / nurture							
Retargeting							
Wrap & measure							

5. Deliverables Checklist

List every asset you'll create and who owns it.

Asset	Owner	Due date
Landing page		
Ad creative (variants)		
Email #1		
Email #2		
Email #3		
Webinar deck / demo		
Lead magnet PDF		
Retargeting ads		
Reporting dashboard		

(Add rows as needed.)

189

6. Metrics Dashboard (define before launch)

Funnel stage	Metric	Target	Actual
Awareness	Ad CTR		
Lead capture	Landing page CVR		
Engagement	Webinar attendance %		
Conversion	Demo requests		
Post-campaign	SQLs / opps		

7. Risk & Mitigation Snapshot

Identify top risks early.

Risk	Likelihood (H/M/L)	Mitigation action

8. AI Assist Prompts

Paste these straight into your custom GPT when you need a boost.

1. **Channel ideas**
2. "Given the 'Innovation Champion' persona and our tagline ‹…›, list three channels with the highest demo-request potential, plus one tactical idea for each."
3. **Email draft**
4. "Write Email #2 in a 3-part sequence for this campaign. Goal: drive webinar sign-ups. Keep tone aligned with our Brand Voice Guide."

5. **Proof point hunt**

6. "Pull compelling stats from our last case study that support Pillar #2 for this campaign copy."

7. **Performance check**

8. "Ad CTR is below target. Suggest two headline tweaks and one image concept to test."

9. Next-Step Commitments

Kick-off meeting date: _____

First asset due: _____

Owner for campaign reporting:

_____

How to Use

1. **Complete Sections 1–2** to anchor the campaign in goals, persona, and messaging.

2. **Brainstorm channels (3)**, draft a **timeline (4)**, and detail **deliverables (5)** to see workload at a glance.

3. **Predefine metrics (6)** so success can't move post-launch.

4. **List risks (7)** to avoid last-minute surprises.

5. **Lean on the AI prompts (8)** anytime you're stuck—it will accelerate copy, ideas, or diagnostics.

6. **Record commitments (9)** so everyone knows who owes what, when.

Fill it, run the plan, measure, iterate—then duplicate for the next campaign.

Part 3 – The Hitchhiker's Guide to Messaging

This book is about building AI assistants for your daily work, not about becoming a marketing expert. There are plenty of brilliant marketing books written by people with decades more experience than I have. But when you're building an AI assistant for marketing, and you aren't a traditional marketer, you need enough foundation to make smart choices.

Thus, my goal here isn't to make you a marketing guru. It's to give you enough understanding to pick the right tool for your situation and know where to dig deeper when needed. Think of this as your practical field guide, not your graduate-level textbook. These extra chapters should provide enough information for you to be dangerous.

10 Messaging Frameworks That Actually Work

People geek out over messaging frameworks, like some people geek out about Pokémon cards. They collect them, read about them, discuss them, but don't really use them. Think of these as merely a helpful structure you can use to relate your needs to the AI assistant to get the output in the form you want.

Here are ten frameworks I can see you using and some pointers on when you might want to use that one.

1. Jobs to Be Done (JTBD)

Think of your customers as hiring managers. In this model, we pretend customers aren't buying products, they are hiring solutions to get a job done. JTBD helps you understand what job your customer is really trying to accomplish, beyond the obvious surface-level features they think they want.

A classic example: People don't buy quarter-inch drill bits because they love drill bits. They buy them because they need quarter-inch holes. But dig deeper, and you'll find they actually need to hang a picture to make their living room feel more like home. That's the real job.

JTBD comes from innovation theory, popularized by Clayton Christensen. It focuses on the functional, emotional, and social outcomes people seek. Instead of building buyer personas around demographics, you map the context, triggers, and struggles that led to their decision.

When it works best:

Product positioning, market discovery, and long-form messaging where you need to tap into genuine customer motivations. Perfect when you're launching something new or trying to break into a crowded market.

When it doesn't: JTBD requires deep customer research and can be a heavy lift for some. If you need quick-turn marketing campaigns or your product is purely transactional, it might be overkill.

The biggest mistake teams make with JTBD? They interview customers about what they want instead of understanding what progress they're trying to make. Ask about the story, not the shopping list.

2. StoryBrand

StoryBrand flips the typical brand narrative. Your brand isn't the hero of the story. Your customer is.

Instead of positioning yourself as the conquering hero, you become the wise guide helping the hero (your customer) overcome a challenge and reach success. It mirrors the classic storytelling arc we've all internalized since childhood.

The framework is beautifully simple: Character faces problem, meets guide, follows plan, achieves success. Your messaging plugs into this familiar structure, making it instantly digestible for your audience.

When it works best: Website copy, brand messaging, sales decks, and elevator pitches where clarity trumps everything else. It's particularly powerful for service-based businesses where trust and guidance are key selling points.

When it doesn't: StoryBrand can feel formulaic, especially in B2B or highly technical contexts. If your audience needs deep technical details or your differentiation is primarily feature-based, the framework might oversimplify your value.

> ### *Pro tip:*
>
> *The hardest part is resisting the urge to make your company the hero. Your ego wants the spotlight, but your bank account needs you to step aside and let the customer shine.*

3. Challenger Sale

Most sales and marketing approaches assume customers know what they need. Challenger messaging assumes they don't.

Born from The Challenger Sale methodology, this framework leads with insight instead of solutions. You teach customers about a problem they didn't fully recognize, reframe their understanding of the market, then guide them to your solution. It's "commercial teaching" at its finest.

The power of Challenger lies in its ability to create urgency around problems that weren't urgent before.

When it works best: B2B sales and marketing where complex buying processes and status quo inertia dominate. Perfect when your solution is innovative or when you're trying to disrupt established patterns.

When it doesn't: Challenger requires strong insight development and can come off as condescending if executed poorly. It's also overkill for simple, transactional purchases where the problem is already well-understood.

The key to Challenger messaging is earning the right to challenge. You need credible insights, not just contrarian opinions. Challenge their thinking, not their intelligence.

4. Golden Circle (Why → How → What)

Simon Sinek's "Start With Why" turned conventional messaging on its head. Instead of starting with what you do, start with why you do it. Technical people need this the most, because they get obsessed with features. Customers don't want features, they want outcomes.

The Golden Circle suggests that inspiring brands communicate from the inside out: Why (purpose) → How (process) → What (product). The logic is simple: People don't buy what you do, they buy why you do it. Emotional resonance and shared values drive buying behavior more than features and benefits.

It's a powerful antidote to commodity positioning. When everyone in your space offers similar features, your "why" becomes the differentiator.

When it works best: Brand storytelling, thought leadership, mission-driven companies, and early-stage startups defining their market position. Particularly effective when your audience has strong values or beliefs.

When it doesn't: Golden Circle can lack specificity for product-focused marketing. Many brands struggle to articulate a credible "why" without sounding manufactured or generic.

Warning: Having a purpose isn't the same as having a profitable business model. Your "why" needs to connect to real customer value, not just feel-good marketing speak.

5. PAS (Problem → Agitate → Solution)

PAS is the framework your grandmother would recognize, even if she doesn't know the name. It's classic, direct-response copywriting at its most fundamental.

Start with a problem your audience recognizes. Stir up the pain, frustration, or cost of not solving it. Then offer your solution as the clear path forward. It's designed to create urgency and emotional resonance quickly.

The "agitate" step is what separates PAS from gentler approaches. You're making the problem impossible to ignore.

When it works best: Conversion-focused messaging where you need to grab attention and move people to action fast. Perfect for ads, landing pages, and email campaigns where immediate response matters.

When it doesn't: PAS can feel manipulative if overdone. Sophisticated audiences may be turned off by aggressive agitation, and it's not ideal for relationship-building or consultative sales processes.

The art of PAS is knowing how hard to push. Agitate enough to create urgency, but not so much that you sound desperate or predatory.

6. AIDA (Attention → Interest → Desire → Action)

AIDA maps the mental journey from passive observer to active buyer. First, grab attention in a crowded marketplace. Then spark genuine interest in your solution. Build desire by connecting to their goals or fears. Finally, prompt clear action. There's <u>a wonderful YouTube clip of AIDA explained by Alec Baldwin in the classic sales movie Glenngary, Glen Ross.</u> *(Strong language warning)*

It's foundational in direct response marketing because it mirrors how people actually make decisions. Even in complex B2B sales, individuals still go through these mental states.

When it works best: Sales letters, email sequences, and product landing pages where guiding user flow matters. It's particularly useful when you control the entire message experience from start to finish.

When it doesn't: AIDA assumes a linear buyer journey, which doesn't always match modern, multi-touchpoint purchase processes. It also works better for individual decision-makers than committee-based buying.

Remember: Attention without interest is just noise. Interest without desire is just entertainment. Desire without action is just frustration. All four elements need to work together.

7. 4 Ps (Promise → Picture → Proof → Push)

The 4 Ps structure builds a complete persuasive argument in four logical steps.

1. Start with a clear promise of value.
2. Paint a vivid picture of what life looks like with your solution.

3. Prove it with evidence, testimonials, or case studies.

4. Then push for action with a clear call-to-action.

It's particularly effective because it combines emotional appeal (picture) with logical validation (proof), addressing both heart and head concerns in a single framework.

When it works best: Landing pages, sales presentations, and marketing messages that need to be both emotionally engaging and credibly backed. Perfect when you have strong social proof to use.

When it doesn't: The 4 Ps assume you have compelling proof points and testimonials to share. If you're launching something new or lack strong case studies, the framework loses much of its power.

The "picture" step is where most teams stumble. Help prospects imagine their improved future state, don't just list benefits. Make it real, specific, and personally relevant.

8. FAB (Features → Advantages → Benefits)

FAB helps bridge the gap between what engineers build and what customers actually care about. Start with a feature (what it is), explain the advantage it provides (how it works), then connect it to the specific benefit for the user (why they care).

It's a translation framework, helping technical teams communicate in customer language. Every feature should ladder up to a meaningful customer outcome.

When it works best: Product descriptions, sales enablement materials, and any context where technical features need translation into user value. Essential for B2B tech companies and complex products.

When it doesn't: FAB can become repetitive and mechanical if overused. It's also less effective for emotional or brand-focused messaging where the technical details matter less than the feeling or status the product provides.

Here's the test: If you can't clearly articulate why a feature matters to your customer's life or business, you probably shouldn't be messaging it in the first place. I remind the founders I consult with all the time, "Features for Features Sake" is just geekery.

9. SPIN Selling (Situation → Problem → Implication → Need-Payoff)

SPIN Selling takes a consultative approach to messaging.

1. Start by understanding the situation.
2. Identify problems within that context
3. Explore the implications of not solving those problems
4. Then position the payoff of your solution.

It's discovery-driven and highly methodical, designed to uncover needs rather than assume them. Power lies in getting prospects to convince themselves of the value, rather than you trying to convince them.

When it works best: Long-form content, B2B sales processes, discovery calls, and customer success case studies. Perfect when the sale requires education and relationship-building over time.

When it doesn't: SPIN is too detailed for short-form content or rapid campaign cycles. It's also better suited to one-on-one conversations than broad marketing messages.

The magic of SPIN is in building a logical flow that leads prospects to their own "aha" moments.

10. Before-After-Bridge

This framework paints a vivid contrast between current frustration and future success, with your solution as the bridge between them. It's transformation-focused and emotionally powerful.

1. **Before**: Here's your painful current state.
2. **After**: Here's your desired future state.
3. **Bridge**: Here's how we get you there.

People need to see themselves in the "before" and "after" scenarios. The key is making both states emotionally resonant and personally relevant.

When it works best: Testimonials, case studies, pitch decks, and any messaging that needs to demonstrate clear transformation or value. Particularly powerful for service-based businesses and personal development products.

When it doesn't: Before-After-Bridge can feel oversimplified for complex solutions or sophisticated audiences. If the transformation isn't dramatic or the timeline is very long, the framework loses impact.

Watch out for: Exaggerating the "before" pain or overselling the "after" outcome. Credibility kills conversion faster than any other messaging mistake.

Choosing Your Framework

Now that you know the tools, how do you pick the right one for your situation?

Start with your primary objective.

Are you trying to educate an unaware market?

Use JTBD or Challenger.

Need to clarify your brand story?

Try StoryBrand or Golden Circle.

Building a high-conversion landing page?

Consider PAS, AIDA, or 4 Ps.

Consider your audience's sophistication. Technical buyers might appreciate FAB or SPIN's consultative approach. Consumer audiences might respond better to Before-After-Bridge or StoryBrand's narrative structure.

Think about your proof points. If you have strong case studies and testimonials, frameworks like 4 Ps or Before-After-Bridge shine. If you're launching something new, JTBD or Golden Circle might serve you better.

Thoroughly Confused Yet?
My best tip for you:

See why I didn't add this chapter in the middle of the book? It's a side quest on the way to building an AI assistant. The confusion is normal probably because you are new to this and want to make the best decision. It'll be freeing to know you can't know the best decision because you aren't your customer.

Your customers don't care which framework you used. They only care whether your message helps them understand how you can make their life better.

You don't need to force a framework that doesn't fit your situation. These are tools, not rules. The best messaging combines elements from multiple frameworks to fit the scenario, rather than rigidly following just one.

You'll need to see how the frameworks you like apply to your business and solutions. As I say that, I sense a certain dread creeping up in your mind. This is dread about wasted effort isn't it? Humans have a need for caloric efficiency and that dread is a long-forgotten survival mechanism.

Remember, your AI Assistant is a machine. ChatGPT is like honey badger, it just don't care.

Pass in your Minimum Viable Company Brief and ask for versions of all 10 messaging frameworks just to see what it comes up with. Get rid of the worst 3. Ask it for which messaging frameworks apply to which buyers' stages, or buyers' personas. Iterate and don't sweat the experimentation. It's cheaper than ever.

Made in the USA
Middletown, DE
24 November 2025

21502630R00116